Springer Tracts in Mechanical Engineering

Springer Tracts in Mechanical Engineering (STME) publishes the latest developments in Mechanical Engineering - quickly, informally and with high quality. The intent is to cover all the main branches of mechanical engineering, both theoretical and applied, including:

- Engineering Design
- Machinery and Machine Elements
- Mechanical Structures and Stress Analysis
- Automotive Engineering
- Engine Technology
- Aerospace Technology and Astronautics
- Nanotechnology and Microengineering
- Control, Robotics, Mechatronics
- MEMS
- Theoretical and Applied Mechanics
- Dynamical Systems, Control
- Fluids Mechanics
- Engineering Thermodynamics, Heat and Mass Transfer
- Manufacturing
- Precision Engineering, Instrumentation, Measurement
- Materials Engineering
- Tribology and Surface Technology

Within the scope of the series are monographs, professional books or graduate textbooks, edited volumes as well as outstanding PhD theses and books purposely devoted to support education in mechanical engineering at graduate and post-graduate levels.

Indexed by SCOPUS, zbMATH, SCImago.

Please check our Lecture Notes in Mechanical Engineering at https://link.springer.com/bookseries/11236 if you are interested in conference proceedings.

To submit a proposal or for further inquiries, please contact the Springer Editor in your region:

Ms. Ella Zhang (China)

Email: ella.zhang@springernature.com

Swati Meherishi (India)

Email: swati.meherishi@springer.com

Dr. Leontina Di Cecco (All other countries)

Email: leontina.dicecco@springer.com

All books published in the series are submitted for consideration in Web of Science.

Dirk Michael Nuernbergk

Overshot Waterwheels

Design Flow and Calculation Examples

Springer

Dirk Michael Nuernbergk
Buchfart, Germany

ISSN 2195-9862 ISSN 2195-9870 (electronic)
Springer Tracts in Mechanical Engineering
ISBN 978-3-031-95384-2 ISBN 978-3-031-95385-9 (eBook)
https://doi.org/10.1007/978-3-031-95385-9

The original submitted manuscript has been translated into English. The translation was done using artificial intelligence. A subsequent revision was performed by the author(s) to further refine the work and to ensure that the translation is appropriate concerning content and scientific correctness. It may, however, read stylistically different from a conventional translation.

Translation from the German language edition: "Wasserräder mit Freihang" by Dirk Michael Nuernbergk, © Verlag Moritz Schäfer GmbH & Co. KG 2007. Published by Verlag Moritz Schäfer GmbH & Co. KG. All Rights Reserved.

This Springer imprint is published by the registered company Springer Nature Switzerland AG
The registered company address is: Gewerbestrasse 11, 6330 Cham, Switzerland

Preface

This book discusses the design and calculation principles of overshot and backshot waterwheels. It is a translation of the second edition of the German version from 2014, with only minor changes and corrections. The proven structure of the book has otherwise been retained.

In recent years, the push for renewable energy generation has intensified, driven by the political will to shift away from fossil and nuclear fuels in favor of renewable energies.

In the field of hydropower, intensive research has been conducted at many institutes, particularly in the area of low head differences, to develop suitable concepts for low-head hydropower sites. Additionally, many small plants, which had been left to decay for six decades, were brought back into operation. Particularly in Germany, the initiative of operators was crucial, as the promotion of small hydropower, unlike other technologies, was not a priority in many federal states.

Since the publication of the first edition, more than 50 sites in Germany have been reactivated with overshot waterwheels.

The overshot waterwheel is an efficient water power machine, known for its excellent performance at both partial and full loads. Its simple construction and high efficiency, compared to other types of waterwheels or even turbines, make it a favorable choice. The installation of overshot waterwheels remains economically viable, even in developed countries. As a result, many smaller hydroelectric sites have been reactivated or newly developed with overshot waterwheels in recent years. The author has personally accompanied the development of ten such sites.

If this book helps achieve the development or reactivation of additional small hydroelectric sites, its goal will be fulfilled.

I would like to express my sincere gratitude to the following individuals for their support: Dr.-Ing. Gerd Treiber, who trusted my skills and manufactured several waterwheels based on the design parameters in this book; Prof. Dr.-Ing. habil. Detlef Aigner from the Technical University in Dresden, who provided valuable measurement data on sharp-edged sluice gates; and, finally, my son Nils, who proofread the English version.

Additionally, I would like to thank Sigrid and Dipl.-Ing. Thomas Ammerschläger, who supported me continuously throughout my entire lifetime.

Dedicated to my grand- and great-grandchildren.

Buchfart, Germany
August 2025

Dirk M. Nuernbergk

Declarations

Competing Interests The author has no competing interests to declare that are relevant to the content of this manuscript.

Contents

List of Figures

List of Tables

Chapter 1
Introduction

1.1 Types of Waterwheels

Waterwheels have existed since antiquity and were independently invented in China and Europe. The first written description of an undershot waterwheel comes from the Roman engineer Vitruvius in his treatises "De Architectura". A very good overview of the historical development of the waterwheel can be found in T. Reynolds [1]. Therefore, the historical development will not be further elaborated upon here.

Figure 1.1 shows the different types of waterwheels. On the one hand, they are differentiated by the shape of the water-filled cavities into cell wheels with bucked- or box-shaped cells and paddle wheels, with flat or bent paddle surfaces that separate the cavities from each other. On the other hand, the wheels are distinguished by the point of water entry:

1. Undershot paddle wheels—the water hits near the lower wheel vertex (e.g. Undershot waterwheel, Poncelet waterwheel in Fig. 1.1a, b and the river wheel not shown here),
2. Mid-shot paddle wheels—the water hits the wheel at the same height or slightly above/below of the wheel axis (e.g. Zuppinger waterwheel and Sagebien waterwheel in Fig. 1.1c, d),
3. mid-shot cell wheels—the water hits the wheel at the same height or slightly above/below of the wheel axis (e.g. Bach's sluice gate wheel in Fig. 1.1e),
4. overshot waterwheels—the water hits in the area of the upper vertex of the wheel (Fig. 1.1f).

Among the overshot wheels, there are also the reverse wheels (which can run forwards and backwards), which have been used in mining. Today, out of the water-

D. M. Nuernbergk, *Overshot Waterwheels*, Springer Tracts in Mechanical Engineering, https://doi.org/10.1007/978-3-031-95385-9_1

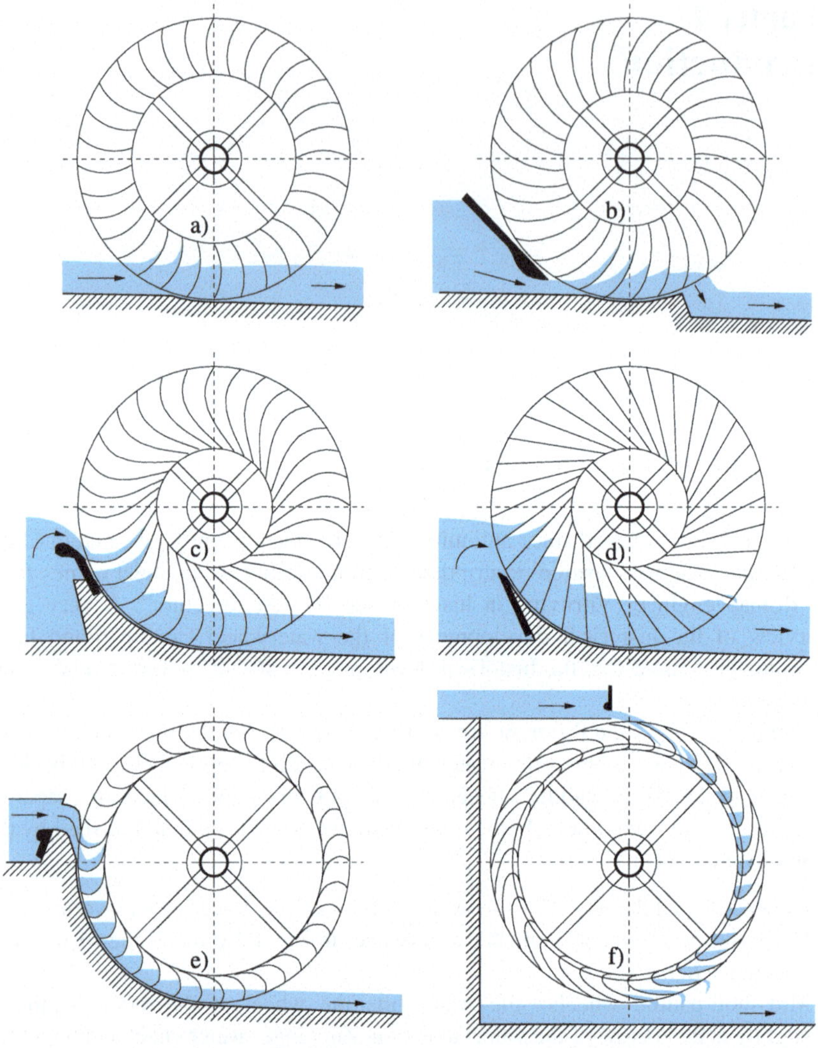

Fig. 1.1 Schematic representation of different waterwheels: (**a**) Undershot waterwheel, (**b**) Poncelet waterwheel, (**c**) Zuppinger waterwheel, (**d**) Sagebien waterwheel, (**e**) waterwheel with sluice inlet according to Bach [2] (backshot waterwheel), (**f**) Overshot waterwheel

wheels described above, mainly the Zuppinger wheel and the overshot waterwheel are built. These two types are mature constructions with high efficiency.[1]

The aim of this volume is to provide the design and calculation basics for the overshot waterwheel. All other paddle and cell wheels are not considered.

[1] Overshot Waterwheel $\eta = 0.8 \ldots 0.87$ [3], Zuppinger waterwheels $\eta = 0.65 \ldots 0.8$ [4, 5].

1.2 The Overshot Waterwheel

In the case of the freely hanging, overshot waterwheel, the water is supplied to the wheel with the speed c_{2A} near the upper vertex of the wheel (see Fig. 1.2). The water is deflected on the cell surface. After the water has come to rest in the cell, it sinks down confined in the cells. This generates a torque M through the weight force of the cell fillings and its distance to the center of the wheel. The torque M and thus also the power $P = \omega \cdot M$ becomes large when the center of gravity of the cell fillings lies far outside. Therefore, the depth of the cells a is kept small towards the center of the wheel.

The overshot wheel generally hangs freely over the tailwater. Wading of the wheel has to be avoided. Even before the lower vertex is reached, premature emptying of the cells begins at the altitude h_1 with the speed c_3. At the height h_2, the cells are completely emptied.

The outflow speed c_4 towards the underwater level cannot be used. The inflow speed c_1 is only significant when the water is supplied to the waterwheel via an open channel or a flume.

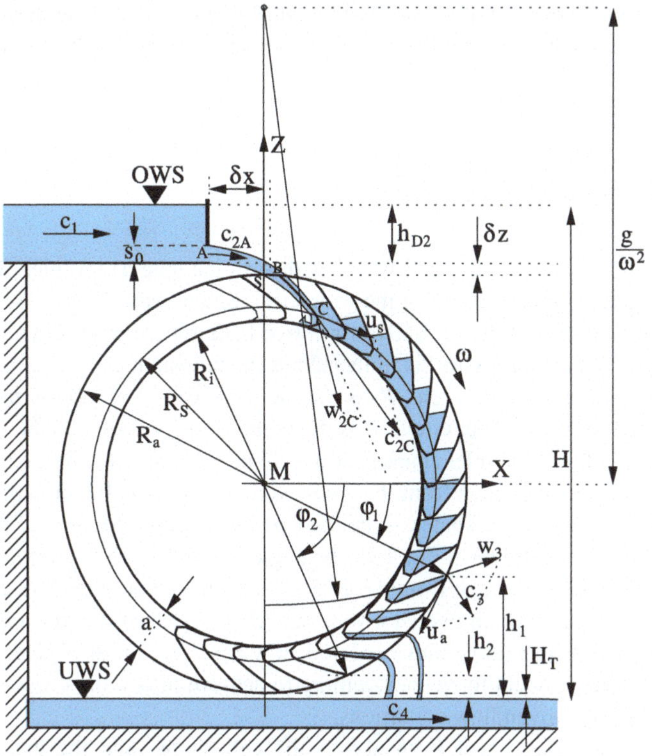

Fig. 1.2 Schematic structure of an overshot cell wheel

The calculation shown in this book of the overshot cell wheel is based on the derivations by Bach [2]. Additional considerations for the numerical calculation of the exit losses are included. Bach has very comprehensively presented the calculation of the overshot waterwheel with an inflow via a channel with sluice gates. In this text, this is extended a bit further and various inflow variants are examined (channels with sluice gates, tongue inlet, open channel/flume).

The entry and exit of the water are analyzed, and the losses are calculated. For this purpose, the shape of the cells must be known. This calculation allows the determination of the efficiency of the waterwheel.

A design flow is presented to provide an easy and systematic calculation methodology of overshot wheels. Chapter 8 is recommended for the impatient reader. There, you the model scaling rules and the design flow can be found with the corresponding references to supplementary images and the other text passages. Two waterwheels are calculated based on the design flow. Both overshot waterwheels were manufactured and now deliver electricity to the main grid.

Furthermore, experimental investigations under laboratory conditions and in real operation are presented. The experimental investigations by Meerwarth [3] are compared with the results of the theoretical calculations. In addition, wading tests by Weidner [6] are presented. This provides a direct comparison between theoretical and experimental results, which later give the reader the opportunity to better evaluate their own calculations. Experiments carried out by the author are presented at the end.

1.2.1 Geometric Dimensions

Figure 1.2 shows an overshot cell wheel with a sluice gate as an inlet. The figure contains all essential definitions of the geometric dimensions. H_T is the immersion depth of the waterwheel. It is negative for a free hanging waterwheel and the value H_T is positive if the waterwheel is immersed in the tailwater.

R_a is the radius of outer circle, R_s is the radius of center of gravity in the cells, and R_i the radius of the inner circle. B is the width of the wheel, a is the rim width, and b_0 is the width of the inlet channel or the sluice gate opening. δx and δz are the distances of the channel edge from the wheel vertex S in X and Z direction.

The origin of the coordinates is defined to be at point M, the center of the waterwheel circle, as shown in Fig. 1.2. The angles φ_1 and φ_2 are the initial and final angles at which the water exit from the cells begins and is completed, respectively.

The task in designing an overshot waterwheel is to determine, for a given flow rate Q (preferably the maximum expected flow rate Q_{max}), at a given head H, the above-defined wheel dimensions, aiming to achieve maximum efficiency of the waterwheel for the given flow conditions.

The entry into the wheel from above, the cell shape, and the exit from the wheel must be carefully determined during design of this type of waterwheel. In this process, it is usually not possible to simultaneously meet all conditions for good entry and exit. An optimum must be found between the impact losses and the premature exit of the water from the cells.

Upon entry into the wheel, the coordinates and the velocities of the water at points A, B, C must be determined for the design. Upon exit, the angles φ_1, φ_2, the velocities w_3 and c_3 and the altitude at which the water exits, must be calculated. With this information, the losses can be inferred and thus the efficiency can be determined. The design of the overshot waterwheel will be dealt with step by step from entry to exit in the following chapters.

1.2.2 Scope of Application

Even today, overshot waterwheels have their justification for existence as water power machines alongside micro turbines. This is due to the favorable efficiency curve of this type of waterwheel ($\eta = 0.75 \ldots 0.85$), the less complex water construction and the simple structure of the overall system. Especially locations with low, highly variable discharge can be developed to a high degree of nominal flow rate (50 days) [3].

The scope of application of overshot waterwheels is shown in Fig. 1.3. The performance shown corresponds to the useful power at the shaft of the wheel. To obtain the theoretical power, these values must be divided by the factor 0.8. The scope of application is limited to a power of about $P = 25$ kW. Above this limit, the

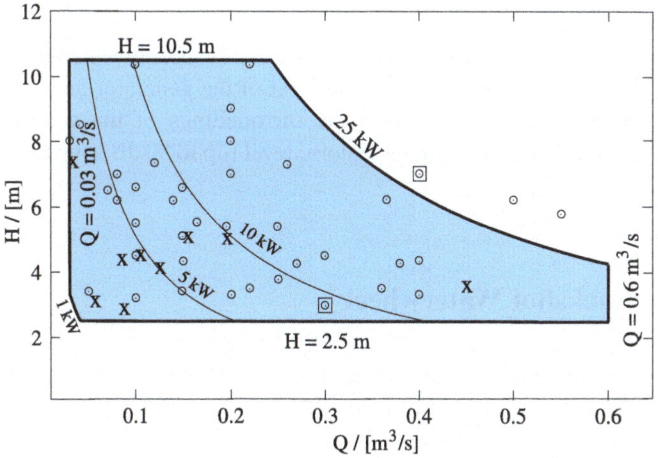

Fig. 1.3 Scope of application of the overshot cell wheel. The circles as data points are taken from the work of Müller [5]. Since Müller almost always gives the minimum and maximum flow, there are two data points shown in the graph for one wheel. The crosses are data points taken from Neumayer [4] (as of 1977). The points framed with a box are the examples calculated in this book and realized at two hydropower sites

use of Francis turbines makes more sense. Only in the special case of refurbishing an old large wheel, this rule is not obeyed.

The flow is limited to a range of $Q = 0.03 \dots 0.6\,\mathrm{m^3/s}$ and the head difference is limited to a range of $H = 2.5 \dots 10.5\,\mathrm{m}$. The specific discharge capacity (specific flow per meter wheel width) is limited to $Q_s = 0.1 \dots 0.25\,\frac{\mathrm{m^3}}{\mathrm{s \cdot m}}$. Only in special cases, wheels are also built for head differences smaller than $H < 2$ m.

The scope of application shown is probably even smaller today ($H \leq 8$ m), as a number of turbines (Pico turbines) are offered that cover the upper range of the head differences. The entire range of head differences and flows shown in Fig. 1.3 can also be covered with cross-flow turbines that have an efficiency of up to $\eta = 0.7 \dots 0.8$.

The efficiency of overshot waterwheels is in the range of $\eta = 0.75 \dots 0.85$ over a wide range of rotational speeds [3] and therefore higher than compared to e.g. cross-flow turbines. The efficiencies of the overshot waterwheel are high, especially at small flow rates, so the machines have a very good partial load behavior.

The choice between the construction of a waterwheel or a turbine therefore depends mainly on the costs of the investment and the location (considerations of monument preservation, existing water construction etc.). Therefore, the use of a waterwheel can only be justified by a comparative cost analysis against other turbo machinery.

The advantages of waterwheels are the very simple hydraulic construction, the easy handling as well as the robustness, the lifespan[2] and the favorable efficiency. However, the disadvantages are the large space requirements, the conditional frost resistance, the high noise level, the low rotational speed n and the resulting large torque M. Rotational speed adjustment can nowadays be done with very compact gearboxes, so that an adjustment to the comparatively high generator speed is no longer a technical difficulty.

The disadvantage of a lack of frost resistance can be circumvented by housing the wheel [5]. There is a inconsiderable waste heat of the generator and the gearbox.[3] This waste heat is then used to heat the surroundings of the waterwheel. The enclosure also muffles the considerable noise level (up to 80 dB sound level close to the wheel).

1.3 The Backshot Waterwheel

The backshot waterwheel is shown in Fig. 1.4. The water is supplied to this waterwheel via a slot entry (see Volume 1 [8]). The waterwheel hangs freely above the tailwater and does not submerge.

---

[2] There are reported cases where over a hundred years of operating life time are proven [4].

[3] 5 . . . 10% of the available mechanical power is converted into heat.

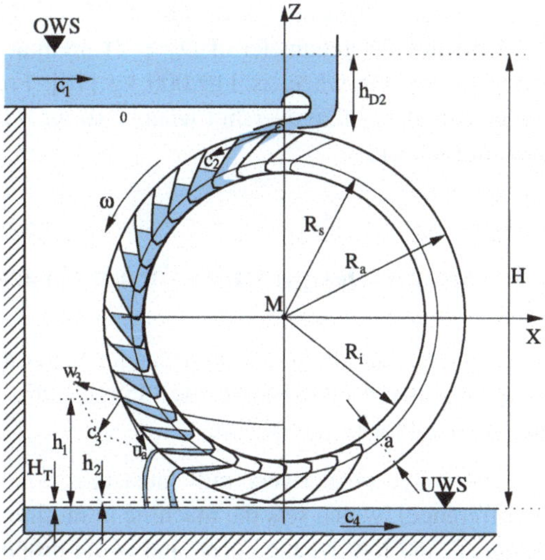

Fig. 1.4 Schematic structure of a backshot cell wheel. To achieve a particularly small wheel diameter, the water is supplied at the upper vertex of the wheel via a slot entry. The flow can be regulated via the shut-off slide with a semicircular profile. This inlet variant was taken from Henne [7]

If one were to add a goiter shaped concrete channel at the height of the wheel center down to the lower vertex, one would directly get Bach's waterwheel with sluice gate entry.

Free-hanging backshot waterwheels have the following disadvantages compared to overshot waterwheels:

1. The waterwheel diameter is generally larger than with the overshot wheel. The large diameter leads to comparatively higher torques.
2. The slot entry is very complex and susceptible to contamination compared to the very simple entry (sluice gate or flume inlet) of the overshot waterwheel.
3. The construction costs are higher due to the slot entry and the larger wheel radius than with the overshot waterwheel.
4. The cells must be ventilated. That means that, either the wheel cells must not be closed towards the inner part and must have exit openings for the air trapped in the cells, or the trapped air must be allowed to escape to the sides of the cells. Side ventilation is only possible with relatively narrow wheels, which limits the scope of application.

The advantages are:

1. The supply of water into the wheel is clearly delineated and avoids large amounts of splash water.
2. The discharge capacity is slightly higher than with the overshot waterwheel.

Backshot waterwheels were used in the head range of $H = 3 \ldots 7\,\mathrm{m}$ in previous times. The wheel diameter is to be chosen according to Müller [9] as $D = 1.3 \ldots 1.4 \cdot H$. The specific discharge capacity is $Q_s = 0.1 \ldots 0.25\ \frac{\mathrm{m}^3}{\mathrm{s \cdot m}}$. The efficiency is slightly lower than that of the overshot waterwheel at $\eta_{WW} = 0.7 \ldots 0.8$.

The largest backshot waterwheel (“Lady Isabella” in Laxey) is located in Scotland and has a diameter of $D_a = 21.35$ m at a head difference of $H = 19.5$ m. The mass of this wheel is 119,000 kg [10]. Free-hanging backshot waterwheels can be calculated like overshot wheels, so we will not go into further detail on the backshot wheel.

1.4 Simple Design Rules for an Overshot Waterwheel

As early as 1825, the following rules for the construction of overshot waterwheels were correctly established in the “Polytechnic Journal” [11]. The unfortunately unknown author is quoted here:

1. „The gauge post: Every mill must have a certain gradient of the water (head difference), which sets the machine in motion. This is indicated and fixed by the upper- and lower-gauge post, and is the property of the mill owner, which must not be diminished in any way. However, the miller must never dam up his water above the gauge post. The better the water difference between the two gauge posts or the gradient is used, the more advantage accrues to the mill. How he achieves this purpose is up to the user, and it would be folly to want to set limits here.

   For example, it is harmless whether the miller pulls his inlet gates a lot or a little, whether it is wide or narrow, high or low, etc. In short, the miller uses his gradient between his two given gauge posts to the best of his knowledge; but takes care that, when high water occurs and exceeds the upper gauge post, the weir gate is opened.
2. The head water (the wet gradient, water weight before the gate) is the water which, when the inlet gate is pulled and the mill is in operation, stands still in front of it, up to the height of the gauge post. Overshot waterwheels should have no (or a small) such water forebay, for it benefits little, but harms much. The quieter the water runs onto the waterwheel, the better it is.

   In mills, which have the clumsy arrangement with a strong water forebay before the gate, the gauge post often stands out 12 inches[4] high above the water level, and the miller should remember that he does not know how to use his gradient well.
3. The fall from the inlet gate to the waterwheel: This often amounts to 4 to 6 inches and is also a waste of the gradient, because as mentioned, the water should flow calmly to the opening through which it falls onto the wheel.
4. The waterwheel: In most mills, the waterwheel is too deep and the cells are spaced too far. Both are disadvantageous. If the cells are too deep (i.e., the rim width a is too large), the water comes too close to the center of the wheel, and

[4] The inch measures were at this time in Germany different from state to state e.g. Bavaria 24.3 mm, Prussia 26.15 mm and Saxony 23.6 mm.

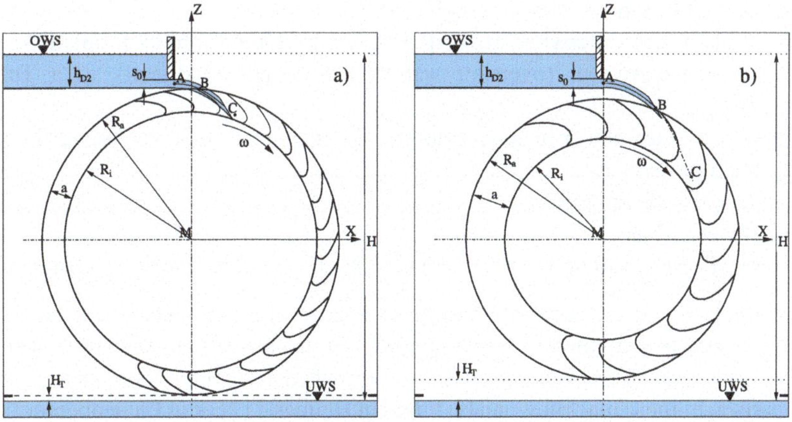

Fig. 1.5 (**a**) Correctly designed overshot waterwheel (with large width B), (**b**) Overshot waterwheel with too deep rim width a, too high distances between neighboring cells, too high free hanging H_T and too large distance from exit point A to working point C (with small width B)

> the weight in the cells of the wheel has a short lever. Therefore, it only acts at the outlet, or when the cell empties. If neighboring cells are too far from each other, the wheel does not take in enough water, resulting in a significant loss of effect. For a waterwheel to be able to take in a sufficient amount of water, the cells it must be narrowly distanced and made as wide as possible (refer Fig. 1.5b)."

With the above said, good overshot waterwheels can be built if these simple rules are followed. In addition, the exit point A should be moved slightly backward and wading of the waterwheel should be avoided by adjusting the wheel diameter to the underwater level so that only at the highest expected flow, the outer ends of the cells touch the underwater level (Fig. 1.5a).

In the following chapters, the individual rules will be discussed in more detail, appropriate advice for the correct design of a waterwheel will be given, and the associated losses will be calculated.

1.5 New Developments

In recent years, there have been a number of developments in the field of hydroelectric machines for low head heights. For example, the following machines were introduced:

- the static-pressure machine (so far, no commercial provider, no efficiency measurement),
- the water-pressure machine (so far, no commercial provider),
- the Steff turbine,

- the lamella turbine (so far, no efficiency measurement),
- the segment-rim waterwheel (overshot waterwheel with stability problems),
- the Turas-waterwheel (overshot waterwheel connected directly to a Turas-gearbox),
- the water-vortex power plant (further development leads to a reaction turbine),
- the hydropower screw,
- the very-low-head turbine,
- the Heber turbine, and
- the overflowed shaft power plant of TU Munich (so far, no commercial provider),

where this list is by no means exhaustive, as only machine sets were mentioned that were technically examined or implemented, and machines that exclusively use the kinetic energy of a flow were omitted. A comprehensive compilation and a current comparison of these machines can be found in Heimerl [12] or in Eichenberger [13]. Turbines, such as the very-low-head and the overflowed shaft power plant, will not be further considered here, as these fall into the category of turbines.

An extensive consideration of the hydropower screw can be found in the volume "Hydropower Screws" [14]. In summary, it can be said that an overshot waterwheel, with the correct design, can be equivalent to a hydropower screw.

Here, it plays a role whether the screw is operated with or without a frequency converter [15]. Compared to a hydroelectric screw with a fixed speed, an overshot waterwheel has better partial load behavior.

When using a frequency converter in the screw system, there is a total loss of 4% due to the converter. This negates the advantage of the higher partial load efficiency of the screw in variable speed operation. Therefore, a decision for one of the concepts can only be made through a cost comparison. Furthermore, heritage conservation aspects and other conditions certainly must be taken into consideration.

For a direct comparison with overshot waterwheels, the water screw, the Steff-Turbine, the segment ring and the Turas waterwheel are interesting concepts.

The segment ring and the Turas waterwheel are modifications of the overshot waterwheel, which only differ from the standard version by constructive peculiarities. For example, in the segment ring waterwheel, the ring is made from individual segments, which can be used for different wheel diameters. However, special attention must be paid to the stability of the wheel construction.

The Turas waterwheel[5] is, like the wheels of a car, mounted on one side. This technique is particularly suitable as a cost-effective solution for small performances. In the hydraulic calculation, these types of waterwheels do not differ from a conventional wheel.

Another concept is the Steff-Turbine (proprietary name). This is actually not a turbine, but a waterwheel with a straight goiter shaped sheet of metal, where the blades curved in the direction of movement are attached to a structure similar to a conveyor belt. The advantages of the arrangement are a very high efficiency at

[5] The Turas is used in tracked vehicles or bucket chains as a drive unit.

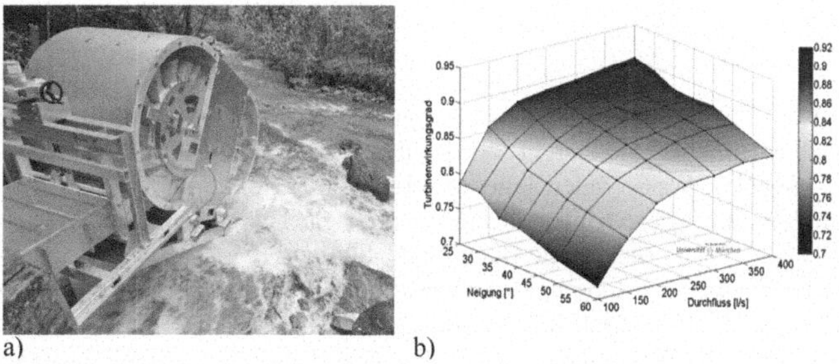

Fig. 1.6 (**a**) Steff-Turbine in operation [16], (**b**) Efficiency curves of the Steff-Turbine, measured at the University of the German Federal Armed Force

Fig. 1.7 (**a**) Romans waterwheel [18], (**b**) Walkers bucket waterwheel [17]

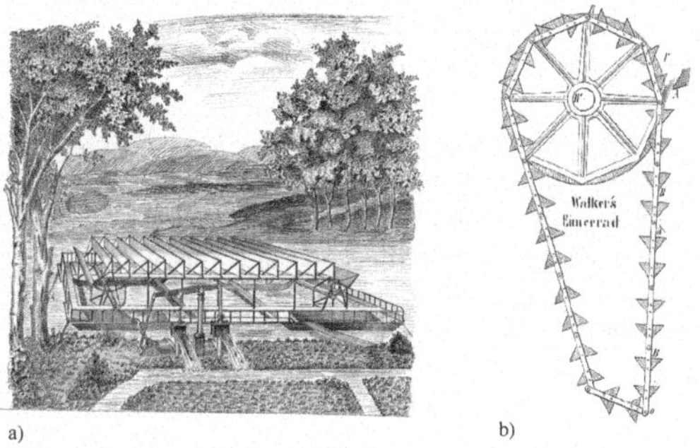

full load (see Fig. 1.6b), a high rotational speed at the driving shaft and that the system can be delivered as a compact unit (see Fig. 1.6a). The compact unit can be adapted to the tailwater level by changing the position. The disadvantages are the high number of moving parts, the lack of operational experience (installation of the first system in 2011 at the Pilgersteg in Räti/Switzerland), so far, insufficient frost resistance and the free suspension of the arrangement. Similar ideas were expressed more than a hundred years ago. Two examples can be found in Fig. 1.7a, b, which date back to 1841 [17] and 1868 [18].

In summary, it can be said that the "new" developments are often based on ideas that were expressed more than a hundred years ago. They thus represent further developments of old concepts. Today, waterwheels must compete with all other mentioned machine concepts. In this context, the area of application, costs, efficiency, construction volume, noise generation and environmental compatibility play an important role.

It is suspected that the overshot waterwheel is very fish-friendly and thus has good environmental compatibility. However, to the author's knowledge, there are no studies on this yet. In addition, overshot waterwheels can be built by hand using this book, as the example of the Obermühle/Niederurff [19] shows.

Therefore, future plant operators can only be advised to obtain offers from different plant concepts and to carry out a cost analysis (construction and plant costs).

References

1. Reynolds TS (1983) Stronger than a hundered men – a history of the vertical water wheel, 1st edn. The Johns Hopkins University Press, Baltimore
2. Bach CV (1886) Die Wasserräder, 1st edn. Verlag von Konrad Wittwer, Stuttgart
3. Meerwarth, KW (1935) Experimentelle und theoretische Untersuchungen am oberschlächtigen Wasserrad. PhD Thesis, Technische Hochschule, Stuttgart
4. Neumayer H, Rempp W, Ruppert J, Schwörer R (1977) Untersuchungen am Wasserradtriebwerk der Kunstmühle W. Seifried KG, Waldkirch/Br
5. Müller W (1999) Die Wasserräder – Berechnung, Konstruktion und Wirkungsgrad, 2nd edn. Verlag Moritz Schäfer, Detmold
6. Weidner CR (1913) Theory and test of an overshot water wheel. Bullet Wisconsin Univ Eng Madison Wisconsin 7(529):157–254
7. Henne H (1903) Die Wasserräder und Turbinen – ihre Berechnung und Konstruktion. Verlag von Bernh. Friedr. Voigt, Leipzig
8. Nuernbergk DM (2005) Wasserräder mit Kropfgerinne – Berechnungsgrundlagen und neue Erkenntnisse, 1st edn. Verlag Moritz Schäfer, Detmold
9. Müller W (1899) Die eisernen Wasserräder – Berechnung, Konstruktion und Wirkungsgrad. Verlag Moritz Schäfer, Detmold
10. Beyrich F (1905) Technische Lehrhefte Maschinenbau Heft 8: Berechnung und Ausführung der Wasserräder. J.M. Gebhardt's Verlag, Leipzig
11. Anonymus (1825) Bessere Benutzung des Aufschlagwassers und des Gefälls bei oberschlächtigen Wasserrädern. Polytech J 17(XXV):103–107
12. Heimerl S, Kohler B (2013) Hydraulische Strömungsmaschinen für kleine Durchflüsse und niedrige Fallhöhen. 43. IWASA, Internationales Wasserbau-Symposium, Aachen 2013 1:19
13. Eichenberger P, Scherrer I, Chapallaz JM (2011) Evaluation von Ultra-Niederdruckkonzepten für Schweizer Flüsse – Innovationen, Eignungskriterien und Erfahrungsberichte. Schlussbericht, Bundesamt für Energie BFE, Eidgenössisches Department für Umwelt, Verkehr, Energie und Kommunikation, Bundesamt für Energie BFE, Forschungsprogramm Wasserkraft, CH-3003 Bern
14. Nuernbergk DM (2012) Wasserkraftschnecken – Berechnung und Entwurf von archimedischen Schnecken als Wasserkraftmaschine, 1st edn. Verlag Moritz Schäfer, Detmold
15. Nuernbergk DM (2023) Hydropower screws – calculation and optimal design of archimedean screws, 1st edn. Springer, Cham
16. WRH Walter Reist Holding AG (2013) Industriestrasse 1, CH-8340 Hinwil/Schweiz. Staffturbine – Linear Power
17. Anonymus (1841) Walker's Eimerrad zur Erreichung des höchsten Effects von einem Wassergefälle. Polytech J 80(C):401–402
18. Anonymus (1868) Roman's schwebendes Ketten-Wasserrad. Polytech J 189(IV):13–14
19. www.obermuehle-niederurff.de (2014)

Chapter 2
Inflow and Outflow Channel

2.1 Open Channel Flow

Most waterwheel systems are run-of-river power plants, which are designed as diversion power plants. An overview of the basics of hydroelectric power plants can be found in Giesecke [1]. The flow movement in an open channel is calculated approximately using the Gauckler-Manning-Strickler formula (abbreviated henceforth as Manning formula). More precise information on the derivation and validity of this equation can be found in Bollrich [2]. The flow velocity and the discharge are then dependent on the gradient[1] I and the dimensions of the channel [2]:

$$c_0 = k_{st} \cdot r_{hy}^{2/3} \cdot I^{1/2}$$

$$Q = A \cdot c_0 = A \cdot k_{st} \cdot r_{hy}^{2/3} \cdot I^{1/2} \tag{2.1}$$

In these formulas, r_{hy} is the hydraulic radius, c_0 is the average flow velocity in the channel, A is the cross-sectional area of the channel, and Q is the channel flow in cubic meters per second.

The value k_{st} is the material-dependent Manning-Strickler coefficient. A compilation of k_{st} values can be found in Table 2.1. The hydraulic radius is defined as the quotient of the cross-sectional area A and the wetted perimeter l_u:

$$r_{hy} = \frac{A}{l_u}$$

[1] Given as a decimal fraction, 1 m gradient on 1000 m corresponds to $I = 0.001$ and not in per mille or percent.

D. M. Nuernbergk, *Overshot Waterwheels*, Springer Tracts in Mechanical Engineering, https://doi.org/10.1007/978-3-031-95385-9_2

Table 2.1 Manning coefficient k_{st} and absolute hydraulic roughness k for open channels. The values are taken from the work of Bollrich [2]

Material	$k_{st}/[\mathrm{m}^{1/3}/\mathrm{s}]$	$k/[\mathrm{mm}]$
Glass, piacryl, polished metal surfaces		0...0.003
Asphalted steel sheet, concrete from steel or vacuum-formwork seamless, smoothed, planed wood joint-free new asbestos cement	90...100	0.1...0.3
Smoothed concrete, smooth plaster, wood planed, well jointed	85...90	0.4...0.6
Well-formed concrete, high cement content	80	0.8
Unplaned wood, concrete pipes	75	1.5
Clinker, ashlar and squared masonry carefully executed, concrete from seamless formwork	70...75	1.5...2.0
Rolled asphalt lining	70	2.0
Rubble masonry carefully executed steel pipe moderately encrusted, concrete unplastered masonry with cement mortar	65...70	3.0
Unplastered concrete, old wooden formwork, plastered masonry, rubble masonry, smooth earth material	60	6.0
Concrete from wooden formwork, old, coarse rubble masonry	55	10
Paved slope, bottom of sand and gravel concrete slabs	45...50	20
Regular earth channels without sediment	40	
Fine gravel, sandy gravel		30
Fine gravel to medium gravel		50
Medium gravel, crushed stone		75
Medium gravel, coarse gravel slightly overgrown earth channels with moderate sediment transport	35	90
Natural riverbed with coarse rubble Earth channels with clumpy clay River floodplain with vegetation	30	
Mountain riverbed with coarse rubble, heavily overgrown earth channels, clumpy earth material	25	
Mountain stream, mountain rivers with strong sediment transport, coarse stone	20	

This is the characteristic dimension of the channel of a certain shape. The channel flow is defined by the hydraulic radius. The hydraulic radius is used as a substitute for the water depth,[2] which cannot be used here.

[2] Only for very wide channels with low water depth does $r_{hy} = h$ apply. The Manning formula is then written:

$$Q = A \cdot c_0 = b_0 \cdot k_{st} \cdot h^{5/3} \cdot I^{1/2}$$

and thus solvable for the water depth h.

For example for a rectangular channel, the hydraulic radius is:

$$r_{hy} = \frac{b_0 \cdot h}{2 \cdot h + b_0}$$

b_0 is the width of the channel and h the water depth. The Manning flow formula is useful in several respects. It can be used to determine the average flow velocity and the water discharge in the channel.

At the same time, it can determine the water depth required for a certain discharge. The discharge capacity of old millrace channels can be determined or new channels can be designed (material selection, trench excavation, channel shape, slope). This way, the demands for certain maximum flow velocities can be met (erosion of the channel).

A disadvantage of the Manning formula is that it cannot be explicitly solved for the water depth. Instead, one needs to apply a graphical or Newton's methods to solve the equation for the water depth.

In the head water and tailwater after the wheel, the water speed must be ensured in such a way that no erosion occurs on the bed or on the side surfaces of the channels. The erosion in a channel is avoided by falling below a mean, critical velocity c_{crit}. The channel does not erode if:

$$c_0 < c_{crit} \tag{2.2}$$

Table 2.2 provides guideline values for the critical velocity for different channel bed materials.

For example, if the water carries sand or suspended particles, then a mean velocity should not be lower than $c = 0.21$ m/s (suspended particles) or $c = 0.42$ m/s (sand) [3]. The average flow velocity in millrace channels should not be chosen below $c = 0.4 \ldots 0.5$ m/s, as sanding-up may be occur.

For trapezoidal channels, the slope angles must also be considered in relation to the channel material. The corresponding values can be found in Table 2.3 as a guideline. The symbol m used in the table and is explained in more detail in Fig. 2.1. The application of the equations given above are shown in the next example.

Example 2.1. ***(Design of an Inflow Channel)*** **An old silted water millrace of 500 m length has to be re-excavated. The soil consists largely of fine gravel. The maximum flow rate planned is** $Q = 80$ l/s. **A trapezoidal channel with a suitable slope angle of the side walls should be used. The dimensions of the channel have to be calculated. In addition, the minimum gradient must be determined and compared with the existing gradient of** 0.5 m **over** 100 m.

From Table 2.3, it can be seen that a slope inclination of 1 : 2 should be chosen for fine gravel. Figure 2.1 defines all the dimensions that determine a trapezoidal channel. The cross section for a trapezoidal channel is:

$$A = b \cdot h + h^2 \cdot m$$

Table 2.2 Critical velocities for erosion in channels. The values are taken from the work of Bollrich [2]

Channel condition	$c_{crit}/[\mathrm{m/s}]$
Colloidal soil	
Loamy sand, loose mud	0.10 ... 0.15
Loose loam	0.15 ... 0.20
Sandy loam	0.40 ... 0.60
Compacted loam, clay, mud	0.70 ... 1.00
Single grain structure	
Silt (grain size: 0.02 ... 0.063 mm)	0.10 ... 0.20
Fine sand (0.02 ... 0.2 mm)	0.20 ... 0.35
Medium sand (0.2 ... 0.63 mm)	0.35 ... 0.45
Coarse sand (0.63 ... 2 mm)	0.45 ... 0.60
Fine gravel (2 ... 6.3 mm)	0.60 ... 0.80
Medium gravel (6.3 ... 20 mm)	0.80 ... 1.25
Coarse gravel (20 ... 63 mm)	1.25 ... 1.60
River gravel, rubble (50 ... 75 mm)	1.70 ... 1.80
River gravel, rubble (75 ... 100 mm)	1.90 ... 2.00
Reinforcements	
Stone fill (100 ... 150 mm)	1.90 ... 3.40
Stone packing (150 ... 200 mm)	2.60 ... 3.40
Stone paving (200 ... 300 mm)	0.10 ... 0.20
Concrete, without sediment movement	4.00
Concrete, with sediment movement	2.50
Cobblestone paving in cement mortar	5.00

Table 2.3 Guideline values for slope inclinations of the channel side walls [2]

Soil type	$1 : m$
Clay/loam	1 : 3
Fine sand soil	1 : 2 ... 1 : 2.5
Coarse sand soil/fine gravel	1 : 2
	1 : 1.5
Rock/concrete	1 : 0.4 ... 1 : 1

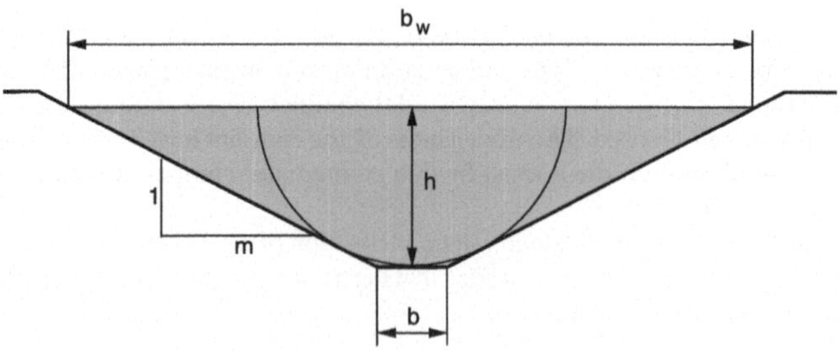

Fig. 2.1 Cross section of the weir channel

Rearranging for b, the result is:

$$b = \frac{A - h^2 \cdot m}{h}$$

The wetted perimeter is:

$$l_u = b + 2 \cdot h \cdot \sqrt{1 + m^2}$$

and by substituting b one gets:

$$l_u = \frac{A - h^2 \cdot m}{h} + 2 \cdot h \cdot \sqrt{1 + m^2}$$

The wetted perimeter is minimal (the hydraulically most favorable cross-section) if

$$\frac{dl_u}{dh} = -\frac{A}{h^2} - m + 2 \cdot \sqrt{1 + m^2} = 0$$

For the water depth h, the equation can be rearranged to yield:

$$h = \sqrt{\frac{A}{2 \cdot \sqrt{1 + m^2} - m}}$$

The necessary flow area results from $A = \frac{Q}{c_{crit}}$. The critical flow velocity should only be reached for the maximum flow rate Q. From Table 2.2, $c_{crit} = 0.6 \ldots 0.8$ m/s is obtained. Thus, $A = \frac{0.08 \text{ m}^3/\text{s}}{0.6 \text{ m/s}} = 0.133 \text{ m}^2$. If this value is inserted into the equations above for the water depth h and the cross section A, the water depth h and the parameter b:

$$h = \sqrt{\frac{A}{2 \cdot \sqrt{1 + m^2} - m}} = \sqrt{\frac{0.133}{2 \cdot \sqrt{5} - 2}} = 0.232 \text{ m}$$

$$b = \frac{A - h^2 \cdot m}{h} = \frac{0.133 - 0.232^2 \cdot 2}{0.232} = 0.109 \text{ m}$$

$$b_w = b + 2 \cdot m \cdot h = 0.109 + 2 \cdot 2 \cdot 0.232 = 1.037 \text{ m}$$

A semicircle with radius h can be drawn into the hydraulically most favorable profile. In Fig. 2.1, the trapezoidal channel with the ratios between b, h and b_w calculated in this example is shown.

The hydraulic radius for a trapezoidal channel is:

$$r_{hy} = \frac{A}{l_u} = \frac{b \cdot h + h^2 \cdot m}{b + 2 \cdot h \cdot \sqrt{1 + m^2}} = \frac{0.133 \text{ m}^2}{1.1465 \text{ m}} = 0.116 \text{ m}$$

It can be easily checked that the hydraulic radius in an optimal trapezoidal channel corresponds to half the water depth:

$$r_{hy} = \frac{h}{2} = \frac{0.232}{2} = 0.116 \text{ m}$$

Now the slope of the channel is checked. From the Manning equation, we get the gradient of the channel bed (with the coefficient $k_{st} = 40$, see Table 2.1):

$$I = \left(\frac{Q}{A \cdot k_{st} \cdot r_{hy}^{2/3}} \right)^2$$

If we insert this value into the equation for the slope, we get:

$$I = \left(\frac{Q}{A \cdot k_{st} \cdot r_{hy}^{2/3}} \right)^2 = \left(\frac{0.08}{0.133 \cdot 40 \cdot 0.116^{2/3}} \right)^2 = 0.003997 \approx 0.004$$

This slope is slightly less than the existing slope of the old canal $I = 0.005$. For this reason, the flow velocity and the discharge capacity is also calculated:

$$c = k_{st} \cdot r_{hy}^{2/3} \cdot I^{1/2} = 40 \cdot 0.116^{2/3} \cdot 0.005^{1/2} = 0.672 \text{ m/s}$$

$$Q = A \cdot c = 0.133 \cdot 0.672 = 0.089 \text{ m}^3/\text{s}$$

The designed channel is already capable of exceeding the maximum flow rate of $Q = 0.08 \text{ m}^3/\text{s}$. The flow velocity for the maximum flow is in the range of the critical velocity for fine gravel. One could remedy this by lining the channel with medium gravel. On the other hand, a change in the channel gradient is also possible. This would result in a gain of 0.5 m over 500 m of head difference. However, a change in the channel gradient is usually not possible (residents, height profile, etc.).

2.2 The Trash Rake

A head loss $h_{V,R}$ results at the inlet rake (see Fig. 2.2) depending on the clear distance a, the shape of the rake bars (bar thickness d), the flow velocity c_0 in front of the rake and the angle of attack α against the flow direction. This is calculated as follows [2]:

$$h_{V,R} = \zeta \cdot \frac{c_0^2}{2 \cdot g} = \beta \cdot \left(\frac{d}{a} \right)^{4/3} \cdot \sin \alpha \cdot \frac{c_0^2}{2 \cdot g} \quad \text{for} \quad c_0 \leq 1 \frac{\text{m}}{\text{s}} \tag{2.3}$$

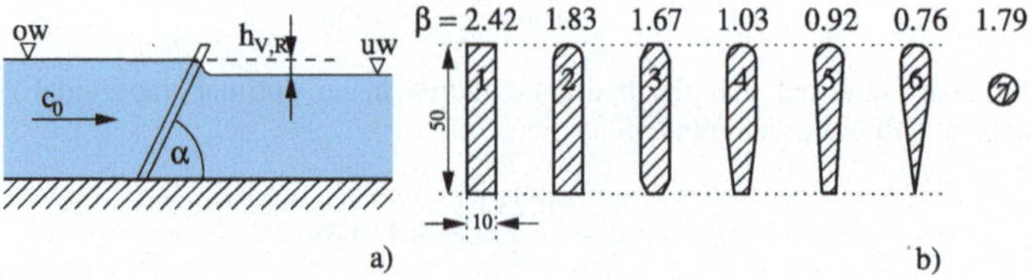

Fig. 2.2 (**a**) Inlet rake, (**b**) form coefficients of different rake profiles (Bollrich [2]). Geometric specifications in mm

The speed c_0 is the inflow speed before the rake, which results from $c_0 = \frac{Q}{A}$, where A is the projection of the rake surface in the direction of flow. Inclined rakes are generally associated with a loss of $h_{v,rake} = 0.1$ m (rarely cleaned) to $h_{v,rake} = 0.05$ m (often cleaned).

Waterwheels tolerate pollution of the water by floating solids better than turbines, as the water enters and exits over the same edge (self-cleaning). A rake cleaning machine is often omitted in waterwheel systems for cost reasons. Usually, overshot waterwheels are operated with a trash rake at the mill site. The debris is already caught at the inflow structure to the mill race at the weir.

The width between the bars can therefore be chosen $a = 60\ldots80$ mm. The round bar profile seems to be a good compromise between the rake losses and the price of the bars. The rake should be placed in front of the inlet structure, inclined and vertically flowed against, to keep the losses small.

2.3 Inlet Structure

The inlet structure close to the mill site requires the construction of the supply to the waterwheel, the feeding of a fish ladder and a bypass channel. The bypass must be able to safely discharge the maximum flow rate when the waterwheel is at a standstill (emergency closure, cold frost weather).

A simple arrangement of the bypass is a spillway, which can be broad-crested or a standard spillway. For the realization of fish ladders, reference can only be made to the respective literature [4], because the requirements can differ from one country to the other.

Upon entry into the inlet structure, the water flows axially and radially on rounded flow paths. Sudden changes to the inlet structure cannot follow these flow paths. This leads to a detachment of the flow and to an undesirable narrowing of the flow cross-section. The water accelerated at the inlet meets slower flowing water in the inlet cross-section, and hydraulic head losses occur.

The flow speed behind the local disturbance is chosen as the reference speed for determining the losses (see Fig. 2.3). The inlet loss head is:

$$h_{V,E} = \zeta \cdot \frac{c_1^2}{2 \cdot g} \tag{2.4}$$

where ζ is the loss coefficient of the inlet structure.

The funnel-shaped inlet has the least losses. The disadvantage is the complex formwork of such a shape. A good variant is an inlet with a circular rounding of the inlet edges. It is possible to gain some centimeters of head compared to a sharp-edged inlet by this means. Since the inflow width b_0 of the channel for waterwheels

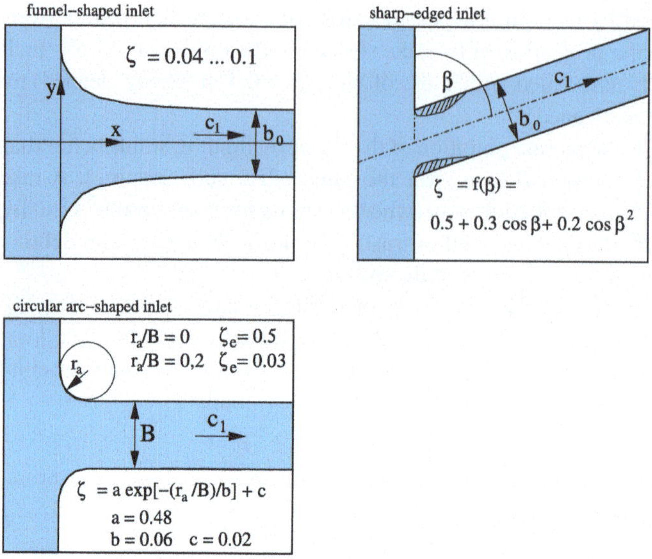

Fig. 2.3 Loss coefficients for different inlet shapes [2]

is in the range of $B = 0.5 \ldots 5$ m, one can choose a favorable entry $r_a/B = 0.2$ ($r_a = 0.1 \ldots 1$ m).[3]

It is also important to avoid cross-sectional changes in the inlet area, otherwise Borda-Carnot losses occur. These result in a narrowing of the flow cross-section, and thus, in losses. This is particularly important when there are installations in the inlet channel (e.g. sluice gates for shutting down the system, pillars).

The guides for sluice gates should therefore be offset into the boundary wall and pillars should be avoided in principle. If such installations are necessary due to site-specific features, they must be given a favorable profile for the water flow.

The effect of cross-sectional changes should not be underestimated. An uneven inflow of the waterwheel results in a decrease in efficiency, as the calculated conditions at the wheel are no longer maintained.

The flow speed in the inflow channel should satisfy $c_1 \leq 1$ m/s. It should also be noted that a semicircle can be drawn into the hydraulically most favorable cross-section. The ratio of width to height is $b = 2.0 \cdot h$ for the hydraulically most favorable cross-section of a rectangular channel (e.g. Fig. 2.4).

A hydraulically favorable profile should be chosen for the short channel section between the inlet structure and entry into the waterwheel.

[3] The given empirical curve $\zeta = a \cdot \exp[-(r_a/B)/b] + c$ applies in the range $r_a/B = 0 \ldots 0.2$.

Fig. 2.4 Inflow channel via an aqueduct (Overshot waterwheel Büschleb Mill, Worbis/Thuringia)

Example 2.2. (Design of the Inlet Channel) **The inlet and the free channel flow to an overshot waterwheel are to be dimensioned for a maximum flow rate of** $Q = 240$ l/s. **The supply of the overshot wheel is realized by a sluice gate with a pressure height of** $h_{D2} = 0.4$ m. **The bypass discharges the water via wall with a weir crest width of** 0.2 m **and is assumed to be a broad-crested weir.**

First, we estimate the wheel width. The specific discharge capacity per unit meter of an overshot wheel is about $Q_s = 0.220\ \mathrm{m^3/(s \cdot m)}$. Thus, a wheel width of $B = Q/Q_s = 1.091\ \mathrm{m} \approx 1.1\ \mathrm{m}$ is needed. The inflow section to the wheel must be about $0.1 \ldots 0.2$ m narrower than the wheel width B. We choose $b_0 = 0.9$ m. The channel in front of the waterwheel should not have any cross-sectional changes.

The water depth in front of the sluice gate is therefore $h = 0.4$ m. Since $2 \cdot h \approx b_0$ applies, the supply channel almost has a hydraulically favorable cross-section. The average flow velocity in front of the sluice is $c_0 = Q/(b_0 \cdot h) = 0.666$ m/s. According to Table 2.2, sediments up to a size of fine gravel can be transported in the approach channel.

Now the bypass is to be dimensioned. The bypass should guide the water beside the overshoot wheel if the water depth is larger than $h = 0.42$ m in front of the sluice gate. The overflow height should be a maximum of $h_{of} = 0.1$ m (measured from the bottom of the bypass wall) to have a freeboard which is not too high (water depth in the channel is then $h = 0.52$ m).

To calculate the overflow, the Poleni equation is used, which yields $Q = C \cdot b \cdot h_{of}^{3/2}$. The coefficient C can be found in Fig. 3.7. A spillover coefficient of $C = 1.47\ \mathrm{m^{1/2}/s}$ is taken for an abrupt inflow edge with $h = 0.1$ m and $w = 0.42$ m (see also Fig. 3.6).

This results in a width of the overflow weir of

$$b = \frac{Q}{C \cdot h^{2/3}} = \frac{0.240}{1.47 \cdot 0.1^{3/2}} = 5.16\ \text{m}$$

By changing the inflow edge according to Fig. 3.7, the discharge height (overflow height) above the weir crown could be slightly reduced, or the weir could be made slightly narrower. A hydraulically favorable shape (high spillover coefficient) with simple manufacturing is desirable.

The maximum freeboard is chosen to be 0.15 m. Therefore, the plate of the sluice gate must be $0.4 + 0.15 = 0.55$ m.

In front of the sluice gate a coarse rake can be arranged, which is vertical arranged to the flow direction, 60° inclined 60° and has a profile made of rounded rods (diameter 10 mm). The rod spacing can be chosen to be very wide at $0.08 \cdots 0.1$ m, as only very coarse driftwood needs to be kept away from the wheel and the sluice.

The calculation of the rake losses under these conditions results in:

$$\begin{aligned} h_{V,R} &= \beta \cdot \left(\frac{d}{a}\right)^{4/3} \cdot \sin\alpha \cdot \frac{c_0^2}{2 \cdot g} \\ &= 1.79 \cdot \left(\frac{10}{100}\right)^{4/3} \cdot \sin 60° \cdot \frac{(0.666\ \text{m/s})^2}{2 \cdot g} \approx 0.002\ \text{m} \end{aligned}$$

The rake losses are negligible if it can be ensured that the rake is cleaned daily by hand. If space is very limited, the bypass can also be realized by a sluice gate or by a siphon overflow.

References

1. Giesecke J, Mosonyi E, Heimerl S (2009) Wasserkraftanlagen – Planung, Bau und Betrieb, 5th edn. Springer Verlag, Berlin
2. Bollrich G, Preißler G (1992) Technische Hydromechanik Band 1, 3rd edn. Verlag für Bauwesen, Berlin
3. Henne H (1903) Die Wasserräder und Turbinen – ihre Berechnung und Konstruktion. Verlag von Bernh. Friedr. Voigt, Leipzig
4. Adam B, Schevers U (2001) Planungshilfen für den Bau funktionsfähiger Fischaufstiegsanlagen, 1st edn. Bibliothek Natur und Wissenschaft, Band 17. VNW-Verlag, Solingen

Chapter 3
Inlet Range and Entry

The water can be supplied by a sluice gate, a tongue inlets, or an open channel. Once the water jet has exited the opening at the speed c_{2A}, the beam falls freely into the wheel (absolute path), until the cells rush through it. At the cell, the jet is divided into a portion falling freely into the cell and a portion striking the back of the blade. The free falling water particles of the jet move on a relative path because of the turning wheel. The other part is forced to move along the plates forming the cells of the wheel. We show how to calculate the absolute and relative path.

The exit velocities c_{2A} for the respective forms of water supply will be determined in the next sections. The exit velocity also determines the flow rate Q that can supply the waterwheel.

Favorable for a high efficiency are a tangential supply of water to the wheel circumference and an entry point B which is near the upper vertex of the wheel. The exit velocity from point A in the channel must be higher than the peripheral speed of the wheel. The circumferential speed at the cell tips should be approximately $u_A = 0.56 \ldots 0.8 \cdot c_{2A}$. To determine the efficiency, one has to determine the losses at the entry into the waterwheel and at the exit of the wheel, which will be calculated in the next chapter.

3.1 Water Supply via a Sluice Gate

We calculate the free outflow from a sluice gate. Figure 3.1 shows the relevant variables.

According to Bollrich [1] the average exit velocity at the exit point A is:

$$c_{2A} = \frac{1}{\sqrt{1 + \frac{\psi \cdot s_0}{h_{D2}}}} \cdot \sqrt{2 \cdot g \cdot h_{D2}} = \varphi \cdot \sqrt{2 \cdot g \cdot h_{D2}} \tag{3.1}$$

D. M. Nuernbergk, *Overshot Waterwheels*, Springer Tracts in Mechanical Engineering, https://doi.org/10.1007/978-3-031-95385-9_3

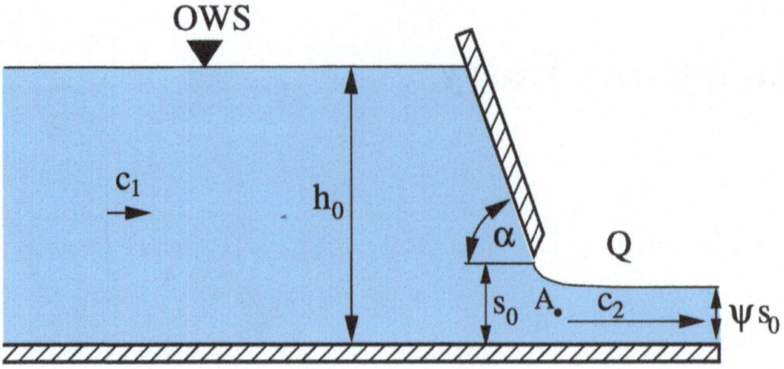

Fig. 3.1 Water entry via a sluice gate

and the flow rate is:

$$Q = b_0 \cdot \psi \cdot s_0 \cdot c_{2A} = \frac{\psi}{\sqrt{1 + \frac{\psi \cdot s_0}{h_{D2}}}} \cdot b_0 \cdot s_0 \cdot \sqrt{2 \cdot g \cdot h_{D2}} \tag{3.2}$$

For the expression $\frac{\psi}{\sqrt{1+\frac{\psi \cdot s_0}{h_0}}}$, technical hydromechanics textbooks introduce the discharge coefficient $\mu_A = \psi \cdot \varphi$. This discharge coefficient depends on the inclination of the sluice gate and the ratio of the pressure head h_0 to sluice gate opening s_0. In addition, the rounding of the sluice in the direction of flow influences the value of μ_A.

For smaller inclination angles α, the discharge coefficient increases, which is desirable, as $\mu_A \approx 1$ should apply (no contraction of the jet, no reduction in jet speed). As an approximation, the following empirically determined formula [1] together with the results of Aigner [2] can be used for a flat, sharp-edged sluice gate:

$$\psi_0 = 1.3 - 0.8 \cdot \sqrt{1 - \left(\frac{\alpha - 205}{220}\right)^2} \quad \text{valid for} \quad s_0/h_0 \approx 0$$

$$\psi = \frac{1}{1 + \left(\frac{1}{\psi_0} - 1\right) \cdot \sqrt{1 - \left(\frac{s_0}{h_0}\right)^{\frac{210^\circ}{\alpha}}}}$$

$$\mu_A = \frac{\psi}{\sqrt{1 + \frac{\psi}{\frac{h_0}{s_0} - 0.5}}} = \psi \cdot \varphi \qquad \text{where} \quad \varphi = \frac{1}{\sqrt{1 + \frac{\psi}{\frac{h_0}{s_0} - 0.5}}}$$

The formulas for the contraction coefficient ψ, the velocity factor φ and the discharge coefficient μ_A apply for attack angles $0^\circ \leq \alpha \leq 180^\circ$ and $s_0/h_0 \leq 1$. Aigner [2] adapted the numerical values in the formula above to his own measurements and measurement values of Gentilini [2].

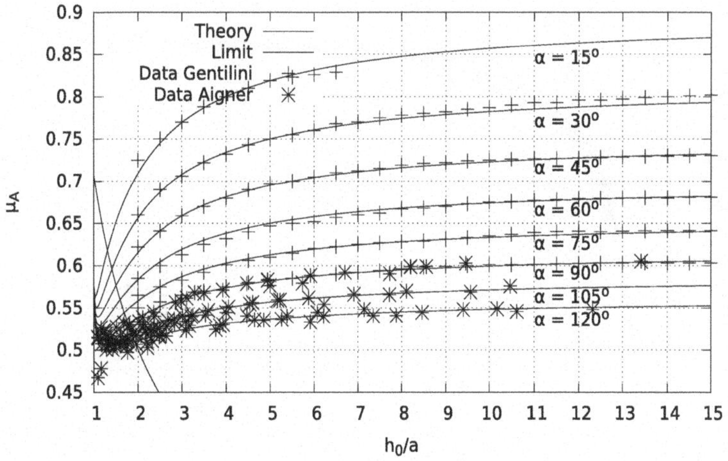

Fig. 3.2 Dimensionless discharge coefficient μ_A for sharp-edged sluices gates as a function of pressure head to opening width h_0/s_0, using the data from Aigner and Gentilini [2]

In Figs. 3.2 and 3.3, the discharge coefficient and the velocity factor are shown together with the boundary condition for the transition between supercritical and subcritical discharge.[1]

It can be seen that the opening width of the sluice gate must not be chosen too large, as otherwise the jet speed or φ varies strongly with increasing opening of the sluice gate. The opening width of the sluice gate should not exceed the ratio $h_0/s_0 = 3 \ldots 4$ as indicated by the Figs. 3.2 and 3.3.

Values of the contraction coefficient of $\psi \approx 0.9 \ldots 0.96$ are obtained for very well-rounded sluice gate entry edges. A rounding is generally recommended, as a larger discharge can be realized with a smaller opening width. The jet width s_0 becomes small and the speed in the jet is high. This is desirable for the water supply of an overshot waterwheel.

At the same time, the hydraulic losses decrease with good rounding. A good guideline for the jet width is $s_0 < 0.125$ m can apply. Otherwise, the required cell division t_z of the waterwheel becomes too large.

[1] The boundary condition results from the condition that the critical depth in the channel is equal to the sluice gate opening. This yields:

$$\mu_{A,crit} \geq \sqrt{\frac{1}{2 \cdot \frac{h_0}{a}}}$$

The boundary conditions results from inserting Eq. (3.2) into the critical water depth of a rectangular channel $h_{crit} = \left(\frac{Q^2}{g \cdot b_0^2}\right)^{1/3} = a$.

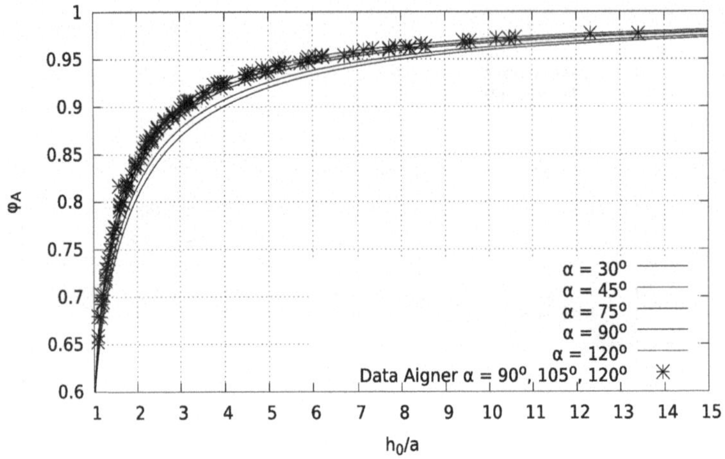

Fig. 3.3 Dimensionless velocity coefficient φ for sharp-edged gates as a function of pressure height to opening width h_0/s_0, using the data from Aigner [2]

The hydraulic losses at a sluice gate are given by:

$$h_{V,e1} = h_{D2} \cdot \left(1 - \varphi^2\right) + \frac{c_1^2}{2 \cdot g} - \psi \cdot s_0$$

The advantage of water entering the overshot waterwheel through a sluice gate is that the exit velocity from the sluice gate remains almost constant if the pressure head h_{D2} is kept constant. The entrance speed towards the wheel are very well maintained. The unavoidable disadvantages are the loss due to the velocity reduction and the risk of contamination by floating debris.

Example 3.1. ***(Design of a Sluice Gate)*** **An inlet of a sharp-edged, vertically arranged sluice gate is to be dimensioned (refer to Fig. 3.4). The maximum flow to be discharged by the overshot waterwheel is $Q_{max} = 0.3\ \frac{\text{m}^3}{\text{s}}$.**

The discharge capacity of the waterwheel determines the channel width in the tail water. One obtains a channel width of $b_0 = Q_{max}/Q_s = 0.3\ \frac{\text{m}^3}{\text{s}}/0.22\ \frac{\text{m}^3}{\text{s}\cdot\text{m}} = 1.364\ \text{m} \approx 1.4\ \text{m}$, if a specific discharge capacity per unit meter of $Q_s = 0.22\ \frac{\text{m}^3}{\text{s}\cdot\text{m}}$ is assumed.

The exit velocity c_{2A} is determined by the pressure height:

$$c_{2A} = \varphi \cdot \sqrt{2 \cdot g \cdot h_{D2}}$$

The maximum opening of the sluice gate should be small compared to the pressure height, $4 \ldots 5 \cdot s_0 \approx h_{D2}$, to avoid water overshooting and pulsation phenomena. One obtains:

$$s_0 = \frac{h_{D2}}{4 \ldots 5} = \frac{Q}{\psi \cdot b_0 \cdot \sqrt{2 \cdot g \cdot h_{D2}}} < 0.125\ \text{m}$$

From these two equations, the pressure height can be determined:

$$h_{D2} = \left(\frac{4\ldots5 \cdot Q}{\sqrt{2 \cdot g} \cdot b_0}\right)^{2/3} = \left(\frac{4\ldots5 \cdot 0.3}{\sqrt{2 \cdot g} \cdot 1.4}\right)^{2/3}$$
$$= 0.33\ldots0.388 \text{ m}$$

The contraction of the jet was neglected $\psi = 1, \varphi = 1$. A value of $h_{D2} = 0.36$ m is chosen. The velocity reduction is set to $\varphi = 1$. Now all sizes of the water supply can be calculated:

$$\psi(\alpha = 90°) = 1.3 - 0.8 \cdot \sqrt{1 - \left(\frac{90 - 205}{220}\right)^2} = 0.618$$

$$\varphi = 1/\sqrt{1 + \psi \cdot s_0/h_{D2}} = 1/\sqrt{1 + 0.618 \cdot 0.125/0.36} = 0.907$$

$$c_{2A} = \varphi \cdot \sqrt{2 \cdot g \cdot h_{D2}} = 0.907 \cdot \sqrt{2 \cdot g \cdot 0.36} = 2.41 \text{ m/s}$$

$$s_0 = \frac{Q}{\psi \cdot b_0 \cdot c_{2A}} = \frac{0.3}{0.618 \cdot 1.4 \cdot 2.41} = 0.144 \text{ m}$$

The opening width of the sluice s_0 does not yet meet the desired condition for the pressure height h_{D2} to ensure the given flow Q, as the contraction of the jet was not taken into account in the calculation. To avoid contraction, one has the option of rounding off the edge of the sluice gate well ($\psi \approx 0.9\ldots0.96$) or increasing the pressure height h_{D2} and repeating the calculation for the new pressure height.

3.2 Water Supply via Bottom Openings

In Fig. 3.4, various possibilities for the outflow of water through an opening from a channel enclosed by fixed walls are shown. The opening is small compared to the

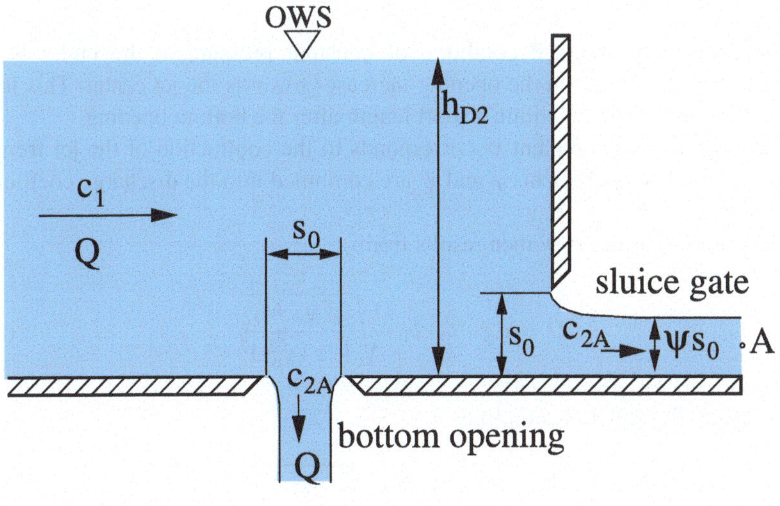

Fig. 3.4 Definitions for the bottom opening and sluice gate

free surface of the channel, which is always the case. Here, we assume a channel with continuous inflow Q with a speed of c_1, so that a steady outflow state is established after a certain time. The upper water level is constant. The outflowing water stream Q and the outflow velocity c_2 from side or bottom openings will now to be calculated.

In the channel, the water level is h_{D2} above a bottom opening with the area $A = s_0 \cdot b_0$. The determining parameters of the bottom opening can be found in Fig. 3.4. The channel width b_0 is not shown. The Bernoulli's energy equation for the case where the pressure at the upper water level is equal to the constant pressure at the outlet opening is:

$$h_{D2} + \frac{c_1^2}{2 \cdot g} = \frac{c_{2A}^2}{2 \cdot g} + h_V$$

The loss height h_V takes into account the influences of internal friction in the jet, friction on the channel walls, and friction in the outlet opening itself. The continuity equation yields $h_{D2} \cdot b_0 \cdot c_1 = s_0 \cdot b_0 \cdot c_{2A}$. If both equations are combined, the exit velocity at point A is:

$$c_{2A} = \sqrt{\frac{2 \cdot g \cdot (h_{D2} - h_V)}{1 - \left(\frac{s_0}{h_{D2}}\right)^2}}$$

The losses h_V can also be written as a velocity reduction factor φ as with the entry over a sluice gate:

$$c_{2A} = \varphi \cdot \sqrt{\frac{2 \cdot g \cdot h_{D2}}{1 - \left(\frac{s_0}{h_{D2}}\right)^2}} \tag{3.3}$$

The previously assumed condition of constant pressure at the outlet is not realistic, as the pressure in the opening increases towards the jet center. This leads to a contraction of the jet within a short length after the bottom opening.

The contraction coefficient ψ corresponds to the contraction of the jet from s_0 to $s_0 \cdot \psi$. The two coefficients φ and ψ are combined into the discharge coefficient $\mu_A = \psi \cdot \varphi$.

The outgoing water flow then results from:

$$Q = \mu_A \cdot s_0 \cdot b_0 \cdot \sqrt{\frac{2 \cdot g \cdot h_{D2}}{1 - (\frac{s_0}{h_{D2}})^2}} \tag{3.4}$$

If $h_{D2} \gg s_0$, the equation simplifies to:

$$Q = \mu_A \cdot s_0 \cdot b_0 \cdot \sqrt{2 \cdot g \cdot h_{D2}}$$

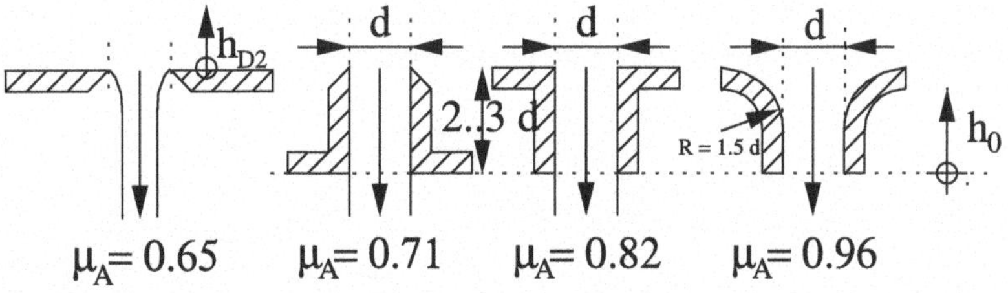

Fig. 3.5 Discharge coefficients for various bottom openings [1]

One obtains a specific velocity at the outlet opening that is related to the outlet area if the velocity is calculated by $c_2 = Q/(b_0 \cdot s_0)$. The average velocity c_2 in the jet is approximately $1/\psi$ of $c_2 = Q/(b_0 \cdot s_0)$ (assuming $s_0/h_{D2} \ll 1$).

In Fig. 3.5, various bottom openings with their discharge coefficients are shown. The steady outflow from a bottom opening with a nozzle (tongue inlet) should be treated just like a bottom opening. The discharge coefficient can be taken from Fig. 3.5. The losses due to friction (velocity reduction) are calculated with the following equation:

$$h_{V,e1} = h_{D2} \cdot \left(1 - \varphi^2\right)$$

Water supply via a bottom opening[2] is not recommended as a water supply of an overshot waterwheel, as the entry angle into the wheel becomes unfavorable. This is because one must allow for a certain jet width. The entry angle is much more favorable when the water is supplied via a sluice gate or an open channel (see also Fig. 3.15).

3.3 Supply via an Open Channel Without Gradient

The supply via an open channel without gradient I is calculated approximately like a broad-crested weir. Detailed explanations on this can be found in Bollrich [1]. On a broad-crested weir, the water flows off with parallel streamlines. The water level drops on the weir crown to the critical depth $h_{gr} = \frac{2}{3} \cdot h_{E,min}$. The critical depth for channels with rectangular cross-section according to Bollrich [1] is:

$$h_{gr} = \sqrt[3]{\frac{Q^2}{g \cdot b_0^2}} \tag{3.5}$$

[2] Also called tongue outlet in the older waterwheel literature.

b_0 is the channel width. If the critical depth is set equal to $h_{gr} = \frac{2}{3} \cdot h_{E,min}$ and the equation is rearranged for the flow rate Q, one obtains

$$Q = \sqrt{g} \cdot b_0 \cdot \left(\frac{2}{3} \cdot h_{E,min}\right)^{3/2} = C \cdot b_0 \cdot h_{E,min}^{3/2}$$

by setting the discharge coefficient $C = \left(\frac{2}{3}\right)^{3/2} \cdot \sqrt{g} = 1.705\ \mathrm{m}^{1/2}/\mathrm{s}$. As with other weirs, the discharge coefficient C depends on the design of the weir and thus on the inflow conditions. This is taken into account by the discharge coefficient of Poleni's formula C depending on both the ratio h/w and the shape of the leading weir edge (see Figs. 3.6 and 3.7).

In Poleni's formula for the broad-crested weir, we now replaced $h_{E,min}$ by h. Poleni's formula then reads:

$$Q = C \cdot b_0 \cdot h^{3/2} \tag{3.6}$$

The exit velocity c_{2A} at point A (refer to Fig. 3.6) will approximately correspond to the critical velocity c_{gr}. This is given by:

$$c_{gr} = c_{2A} = \sqrt{g \cdot h_{gr}} \tag{3.7}$$

This speed determines the flow rate Q and the exit velocity c_{2A} at point A from the channel.

When designing an overshot wheel with channel inflow, a maximum specific flow capacity of $Q_s = 0.2 \ldots 0.22\ \mathrm{m}^3/(\mathrm{s} \cdot \mathrm{m})$ is assumed. With the maximum flow rate, the width of the channel b_0 and thus the width of the waterwheel $B = b_0 + 0.1 \ldots 0.2$ m is determined. The exit velocity of an arrangement like in Fig. 3.6 is still too low and the water jet is too thick to supply an overshot waterwheel.

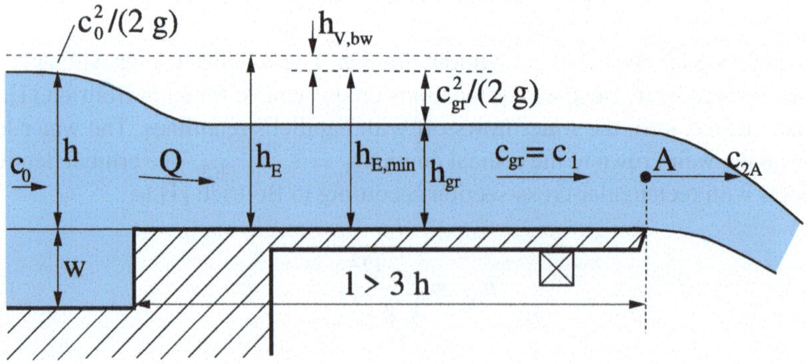

Fig. 3.6 Broad-crested weir with abrupt inflow edge

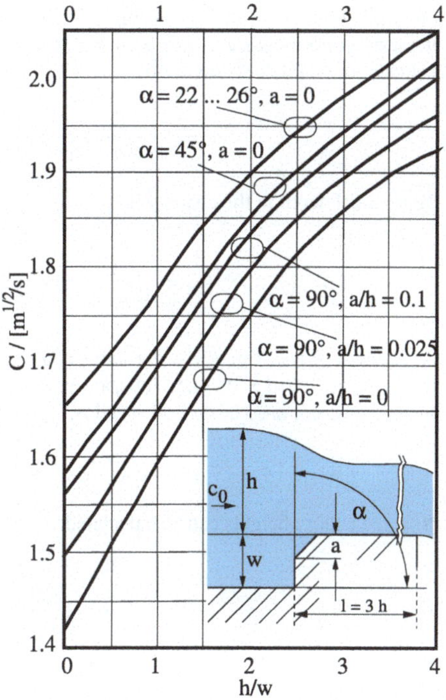

Fig. 3.7 Coefficients of the Poleni's formula for broad-crested weirs (Bollrich [1])

The calculation method presented here only applies for a horizontally arranged channel without gradient. If the channel has a gradient, the design for a chute must be used, which is explained in more detail in the next section. The following example illustrates how to apply the above equations and figures.

Example 3.2. ***(Inflow with Open Channel Without Gradient)*** **A horizontally arranged channel inflow is to be dimensioned ($\alpha = 90°, a = 0$). All information refers to Figs. 3.6 and 3.7. The maximum flow rate should be $Q_{max} = 0.2\ \frac{m^3}{s}$. The water depth in the approaching water channel at maximum flow is $h_{Ch} = h + w = 0.5$ m. The water channel and the subsequent channel are of the same width ($b_0 = b_{Ch}$). The weir crest of the broad-crested overflow with rectangular leading edge should have a length of $l = 3$ m. All relevant data b_0, h, w, c_{gr}, h_{gr}, c_{2A}, $h_{V,bw}$ and c_0 of the channel must be calculated. The calculation process is to be repeated for half the flow.**

The waterwheel's discharge capacity determines the channel width. Assuming a specific discharge capacity of $Q_s = 0.2\ \frac{m^3}{s \cdot m}$, one obtains a channel width of $b_0 = Q_{max}/Q_s = 0.2\ \frac{m^3}{s}/0.2\ \frac{m^3}{s \cdot m} = 1.0$ m. The wheel width is then $B = b_0 + 0.2$ m $= 1.2$ m. The overflow height h is then determined from Poleni's formula. The theoretical value of $C = 1.7\ m^{1/2}/s$ is used as the starting value. One calculates:

$$h = \left(\frac{Q}{C \cdot b_0}\right)^{2/3} = \left(\frac{0.2}{1.7 \cdot 1.0}\right)^{2/3} = 0.24 \text{ m}$$

With the water depth in the water ditch of $h_{Ch} = 0.5$ m, a weir height of $w = h_{Ch} - h = 0.26$ m results. Now, a correction of the coefficient C is necessary. For the value $h/w \approx 1$

one finds $C = 1.58\ \mathrm{m}^{1/2}/\mathrm{s}$ in Fig. 3.7. The overflow height is $h = 0.252$ m. A further iteration is not necessary. The weir height is thus $w = 0.248\ \mathrm{m} \approx 0.25$ m. The inflow velocity in the ditch results from:

$$c_0 = \frac{Q}{(h+w)\cdot b_0} = \frac{0.2}{0.5\cdot 1.0} = 0.4\ \mathrm{m/s}$$

The critical depth in the channel is:

$$h_{gr} = \sqrt[3]{\frac{Q^2}{g\cdot b_0^2}} = \sqrt[3]{\frac{0.2^2}{g\cdot 1.0^2}} = 0.16\ \mathrm{m}$$

At a length of $l = 3$ m, the water in the channel has a mass of $m = \varrho \cdot V = 999\ \mathrm{kg/m^3} \cdot 0.16\ \mathrm{m}\cdot 3\ \mathrm{m}\cdot 1.0\ \mathrm{m} = 480\ \mathrm{kg}$. Thus, a weight force of $m\cdot g = 4710$ Nm has to be supported by the construction. The critical speed is:

$$c_{2A} \approx c_{gr} = \sqrt{g\cdot h_{gr}} = \sqrt{9.81\cdot 0.16} = 1.253\ \mathrm{m/s}$$

Now the energy heights can be calculated:

$$h_E = h + \frac{c_0^2}{2\cdot g} = 0.252 + \frac{0.4^2}{2\cdot 9.81} = 0.26\ \mathrm{m}$$

$$h_{E,min} = h_{gr} + \frac{c_{gr}^2}{2\cdot g} = \frac{3}{2}\cdot h_{gr} = \frac{3}{2}\cdot 0.16 = 0.24\ \mathrm{m}$$

which yields a loss of:

$$h_{V,bw} = h_E - h_{E,min} = 0.26 - 0.24 = 0.02\ \mathrm{m}$$

This loss can be reduced by improving the leading edge of the weir (rounding or e.g. $\alpha = 45°$). For half the flow $Q = 0.1\ \frac{\mathrm{m}^3}{\mathrm{s}}$, if sill height is kept to $w = 0.25$ m one obtains:

$$\begin{aligned}
h &= 0.159\ \mathrm{m} \quad (C = 1.58\ \mathrm{m}^{1/2}/\mathrm{s\ assumed})\\
c_0 &= 0.244\ \mathrm{m/s}\\
h_{gr} &= 0.1\ \mathrm{m}\\
c_{gr} &= c_{2A} = 0.99\ \mathrm{m/s}\\
h_E &= 0.1625\ \mathrm{m}\\
h_{E,min} &= 0.15\ \mathrm{m}\\
h_{V,bw} &= 0.0125\ \mathrm{m}
\end{aligned}$$

The upstream water level has dropped by 0.09 m compared to the maximum flow. This manifests itself as a loss of head due to the change of the water level in the upstream area of $h_{V,OWS} = 0.09$ m. The total loss $h_V = h_{V,bw} + h_{V,OWS}$ already occurs before the water has entered the wheel. If the upstream water level were regulated by a sluice gate, this loss could be compensated. It is also unfavorable that the inlet velocity has dropped to $c_{2A} = 0.99$ m/s. This causes a shock on the cell back, as will be shown later.

Fig. 3.8 Overshot waterwheel in construction with inflow via an open channel supply. The channel can be opened downwards with a bypass flap. To prevent the wheel from running backwards when the bypass flap is open, a wooden wall was inserted

The equations above only apply to a channel without slope I. However, they illustrate the influence of the channel entry on the conditions on the broad-crowned weir crest very well. As the example shows, a straight channel section as a supply is not well suited to an overshot waterwheel (see Fig. 3.8).

The reason for this is that the outlet velocity c_{2A} becomes too low to meet $u_A = 0.56 \ldots 0.82 \cdot c_{2A}$ as a condition derived later.[3] The inflow velocity can be increased using a slope of the channel, as will be shown in the next section. The starting point for these calculations are the explanations given in this section.

3.4 Supply via an Open Channel with Slope

To increase the entry velocity into the waterwheel, the horizontal channel piece of the broad-crowned weir is extended by a long rectangular, inclined chute. One then obtains an arrangement that corresponds to a measuring channel.[4]

[3] The circumferential velocity has to be chosen equal to half of the inflow speed.

[4] This arrangement is also referred to as a flume.

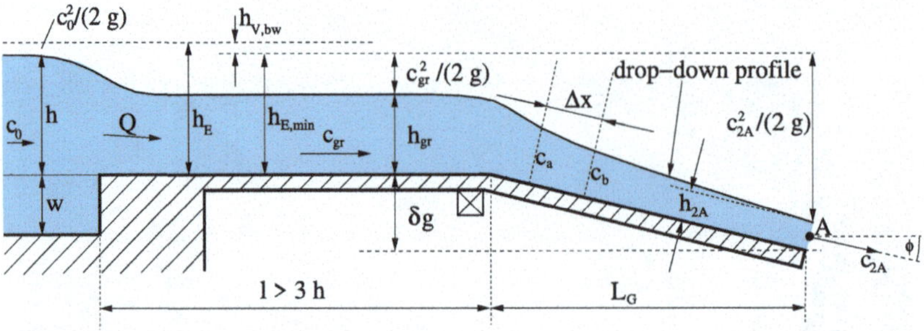

Fig. 3.9 Broad-crowned weir, extended by a chute

Table 3.1 Differentiation of flow states (Bollrich [1])

Flow state	Water depth	Speed	Slope
Subcritical flow condition	$h > h_{gr}$	$c < c_{gr}$	$I < I_{gr}$
Critical state	$h = h_{gr}$	$c = c_{gr}$	$I = I_{gr}$
Supercritical flow condition	$h < h_{gr}$	$c > c_{gr}$	$I > I_{gr}$

This arrangement is used to measure the flow in free-surface channels. The basic arrangement can be found in Fig. 3.9.

For the calculation of this type of measuring channels, there is the freely available software "WinFlume", which can be found under [3]. There, we find additional literature references on the dimensioning of measuring channels (flumes).

The flow conditions on the horizontal part of the weir back are well known, as the critical water depth and critical speed occur. In the inclined range of the channel, however, the water is accelerated and the water level declines. There is a supercritical flow in the chute, because the water has already reached the critical speed on the horizontal weir crest.

In the case of supercritical flow, the flow velocity is higher than the critical speed and the water depth is less than the critical depth (see Table 3.1). The limiting depth for a rectangular channel:

$$h_{gr} = \sqrt[3]{\frac{Q^2}{g \cdot b_0^2}}$$

has to be compared with the water depth h determined from Manning's formula:

$$Q = A \cdot c_1 = b_0 \cdot h \cdot k_{st} \cdot r_{hy}^{2/3} \cdot I^{1/2} \quad \text{with}$$

$$r_{hy} = \frac{A}{l_u} = \frac{b_0 \cdot h}{2 \cdot h + b_0} \quad \text{(rectangular channel)}$$

for a given discharge Q. Manning's coefficients k_{st} are taken from Table 2.1. For this, the key curve must be determined iteratively, as Manning's equation cannot be explicitly solved for the water depth.[5] Since the critical depth is already reached on the weir crest, the water depth in the subsequent channel (chute) continues to decrease and one always reaches supercritical flow conditions with this arrangement.

The speed c_1 is then easy to determine with the equation:

$$c_1 = k_{st} \cdot r_{hy}^{2/3} \cdot I^{1/2}$$

given the known water depth h. The exit speed from the channel should approximately be $c_{2A} \approx c_1$. This value has to be compared to the critical speed:

$$c_{gr} = \sqrt{g \cdot h_{gr}}$$

for the rectangular channel. An example should clarify the calculation process.

Example 3.3. ***(Inflow over an Open, Inclined Channel (Manning's Equation))*** **An inclined steel channel inflow is to be dimensioned. The maximum achievable flow rate is** $Q_{max} = 0.2\ \frac{\text{m}^3}{\text{s}}$. **The channel has a length of** $l = 30$ m **and is lowered by** 60 cm **over this length. All relevant data** b_0, h, c_{gr}, h_{gr}, c_{2A} **and** c_1 **of the channel must be calculated.**

The channel width is $b_0 = Q_{max}/Q_s = 0.2\ \frac{\text{m}^3}{\text{s}}/0.2\ \frac{\text{m}^3}{\text{s}\cdot\text{m}} = 1.0$ m. The slope is $I = 0.6\ \text{m}/30\ \text{m} = 0.02$. According to Table 2.1 Manning's coefficient $k_{st} = 100$. With the starting value:

$$h_1 = \left(\frac{Q}{b_0 \cdot k_{st} \cdot I^{1/2}}\right)^{3/5} = \left(\frac{0.2}{1 \cdot 100 \cdot 0.02^{1/2}}\right)^{3/5} = 0.0777\ \text{m}$$

Newton's method yields:

$$F(h_i) = Q - k \cdot \frac{h_i^{5/3}}{(2 \cdot h_i + b_0)^{2/3}}$$

$$F'(h_i) = -k \cdot \frac{5/3 \cdot h_i^{2/3} \cdot (2 \cdot h_i + b_0)^{2/3} - 4/3 \cdot h_i^{5/3} \cdot (2 \cdot h_i + b_0)^{-1/3}}{(2 \cdot h_i + b_0)^{4/3}}$$

$$k = b_0^{5/3} \cdot kst \cdot \sqrt{I}$$

$$h_{i+1} = h_i - \frac{F(h_i)}{F'(h_i)}$$

[5] A simple solution (starting value for the iteration) for the water depth for a given flow rate Q results in a very wide channel, where $h \ll b_0$ should apply. The hydraulic radius is then equal to the water depth h. Thus, the flow rate from Manning's equation (2.1) yields:

$$Q = b_0 \cdot k_{st} \cdot h^{5/3} \cdot I^{1/2}$$

This equation can be solved for h and one obtains:

$$h = \left(\frac{Q}{b_0 \cdot k_{st} \cdot I^{1/2}}\right)^{3/5}$$

Just four iterations yield a result accurate to ten places:

$$h_1 = 0.07767996097 \text{ m}$$

$$h_2 = 0.08265798325 \text{ m}$$

$$h_3 = 0.08257767136 \text{ m}$$

$$h_4 = 0.08257765104 \text{ m}$$

$$h_5 = 0.08257765104 \text{ m}$$

where only an accuracy of four decimal places is required. The rapid convergence is due the very good starting value for the calculation. The speed in the channel can now be determined with the water depth:

$$c_1 = k_{st} \cdot r_{hy}^{2/3} \cdot I^{1/2} = 100 \cdot \left(\frac{1 \cdot 0.08258}{2 \cdot 0.08258 + 1} \right)^{2/3} \cdot 0.02^{1/2} = 2.422 \text{ m/s}$$

This corresponds to the exit speed $c_1 = c_{2A}$. A comparison with the critical depth for the rectangular channel:

$$h_{gr} = \sqrt[3]{\frac{Q^2}{g \cdot b_0^2}} = \sqrt[3]{\frac{0.2^2}{9.81 \cdot 1^2}} = 0.1597 \text{ m}$$

and the critical speed:

$$c_{gr} = \sqrt{g \cdot h_{gr}} = \sqrt{9.81 \cdot 0.1597} = 1.252 \text{ m/s}$$

shows that supercritical flow is present in the channel. An arrangement of the channel without inclination (broad-crested weir) would result in too high a water depth and too low an exit speed c_{2A} at the exit of the chute. The channel material should have a high Manning's coefficient k_{st} (steel instead of wood) in order to achieve the highest exit speed c_{2A} with a low inclination of the channel. So far, however, we do not know the point (and the necessary length) at which a stationary water depth h (according to the result of Manning's formula) has been established. This question will be answered next.

When water enters the chute, a transition from subcritical to supercritical flow occurs. A stationary non-uniform flow process occurs, as the flow is accelerated until the delay due to friction losses and the acceleration cancel each other out. It is assumed that with a sufficiently long chute, the velocity c_1 and water depth h calculated according to Manning's formula (2.1) will be reached and then remain constant.

The necessary length L_G of the chute can be determined via the calculation of the water surface line (drop-down profile), taking into account the losses due to friction. In Bollrich [1], one can find the dx and dh methods for approximate calculation of the water surface line (Bresse curves).

For the idealized case of no losses in the chute (valid for short flow distances), a simple relationship for the water depth can be derived. From Bernoulli's equation

$$\delta g + \frac{c_{gr}^2}{2 \cdot g} + h_{gr} = \frac{c_{2A}^2}{2 \cdot g} + h_{2A} \tag{3.8}$$

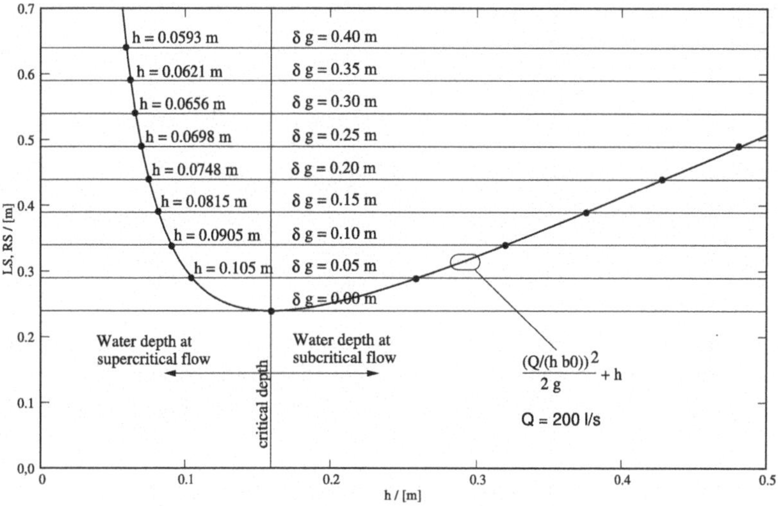

Fig. 3.10 Graphical solution of the cubic equation for the water depth

and the continuity equation

$$Q = b_0 \cdot h_{2A} \cdot c_{2A}$$

one obtains the cubic equation for the water depth, which can be solved graphically.[6] This is shown in Fig. 3.10 for a flow rate of $Q = 0.2\ \text{m}^3/\text{s}$. A weakness of the equation above is that the slope I in the chute and the length L_G for the flow process do not play a role. That is because the friction in the chute is neglected.

The determined water depths are too low. (Compare to Fig. 3.13, in which results that consider friction are compared with those without friction in the chute.) Thus, for very long chutes, one obtains erroneous results, as the friction on the chute walls is neglected.

Therefore, for a more accurate determination of the water depth in the chute, we use the step-by-step calculation of the water surface line. The Δx method is a suitable method for this purpose. It is assumed that there are two cross sections a and b in the chute (see Fig. 3.9) with a small distance Δx. The differential equation of the drop-down profile of the water surface (derivation e.g. at Bollrich [1]):

$$\frac{dh}{dx} = (I_0 - I_E) - \frac{1}{2 \cdot g} \cdot \frac{d(c^2)}{dx}$$

[6] To do this, plot the left and the right side of the equation on a graph and determine the intersections as solutions. One obtains two solutions, one for the subcritical and one for the supercritical flow.

can be rewritten to a difference equation:

$$0 = \Delta h - I_0 \cdot \Delta x + I_E \cdot \Delta x + \frac{c_b^2 - c_a^2}{2 \cdot g}$$

where the slope of the energy line I_E is replaced using Manning's equation (2.1) by:

$$I_E = \frac{c_m^2}{k_{st}^2 \cdot r_{hy}^{4/3}}$$

I_0 is the slope of the channel (chute), c_a and c_b are the velocities at the two cross-sections a and b in the channel. The final equation using average values for the hydraulic radius $r_{hy,m}$ and the water velocity c_m for a rectangular channel are then:

$$0 = \Delta h - I_0 \cdot \Delta x + \frac{c_m^2}{k_{st}^2 \cdot r_{hy,m}^{4/3}} \cdot \Delta x + \frac{c_m}{g} \cdot \Delta c \quad \text{where}$$

$$c_m = \frac{c_a + c_b}{2} \qquad c_a = \frac{Q}{h_a \cdot b_0} \qquad c_b = \frac{Q}{(h_b - \Delta h) \cdot b_0}$$

$$\Delta c = c_b - c_a$$

$$r_{hy,m} = \frac{r_{hy,a} + r_{hy,b}}{2} \quad \text{and} \quad r_{hy,a} = \frac{b_0 \cdot h}{2 \cdot h + b_0} \quad \text{or} \quad r_{hy,b} = \frac{b_0 \cdot (h - \Delta h)}{2 \cdot (h - \Delta h) + b_0}$$

This equation can be rearranged for Δx:

$$\Delta x = \frac{\Delta h + \frac{c_m}{g} \cdot \Delta c}{I_0 - \frac{c_m^2}{k_{st}^2 \cdot r_{hy,m}^{4/3}}} \tag{3.9}$$

As a starting value for the calculation at point a, the critical depth $h_a = h_{Gr}$ is known in the horizontally running channel piece. By specifying small Δh values, a Δx can be determined. Here, Δx is the distance from the point[7] where the flow changes its behavior to the point where the water depth $h_{gr} - \Delta h$ is reached.

The direction of the calculation is heading downstream. By trying several Δh, the water depth $h_{2A} = h_{Gr} - \Delta h$ at the end of the channel of length $\Delta x = L_G$ is obtained.

It should be noted that smaller water depths than calculated with Manning's equation do not occur in the chute (see previous example). For the water depth calculated with Manning's equation, the value of Δx becomes infinitely large.

The calculation results of the Δx method are shown in the following Figs. 3.11, 3.12, and 3.13. The calculations were carried out for the flow rate $Q = 0.2$ m^3/s.

[7] The transition point between the channel piece without slope and the section with inclination.

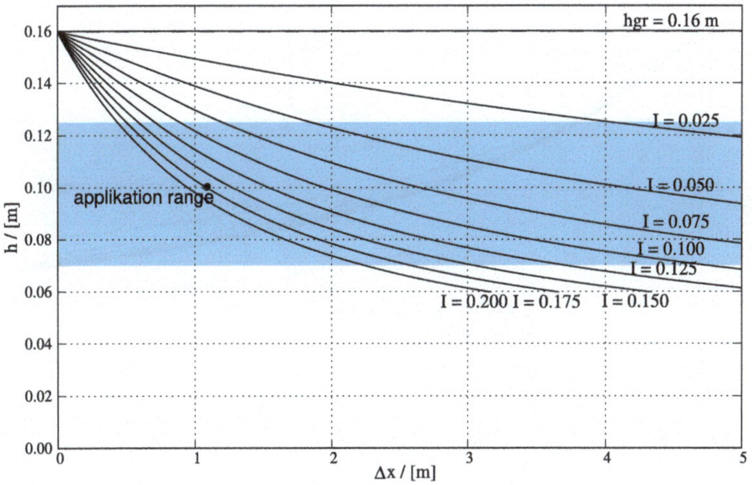

Fig. 3.11 Water depth h plotted versus the length Δx. $k_{st} = 100$, $Q = 200\,\mathrm{l/s}$

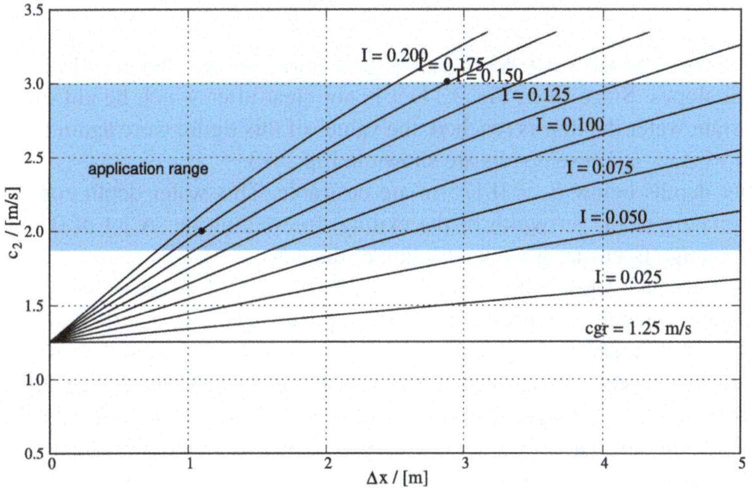

Fig. 3.12 Velocity plotted versus the length Δx. $k_{st} = 100$, $Q = 200\,\mathrm{l/s}$

Manning's coefficient of the rectangular channel was set for steel with $k_{st} = 100\,\mathrm{m}^{1/3}/\mathrm{s}$.

The width of the channel was $b_0 = 1$ m. The water depths for different flows can be found in Fig. 3.11. The exit velocity of the water from the channel can be calculated:

$$c_{2A} = \frac{Q}{h_{2A} \cdot b_0} \tag{3.10}$$

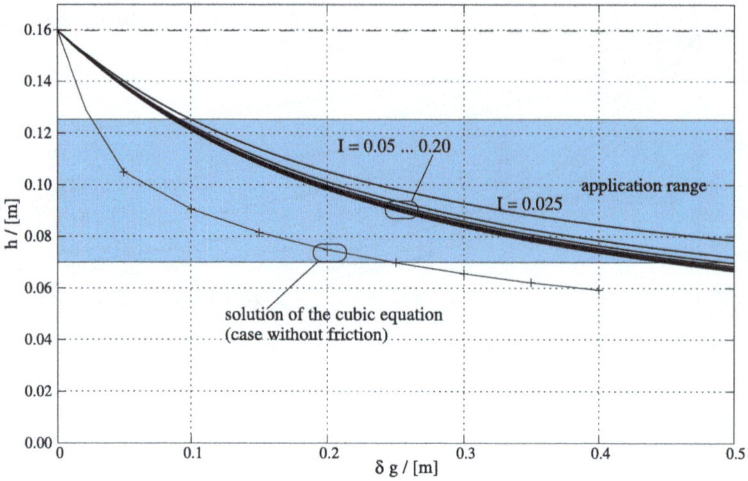

Fig. 3.13 Water depth h versus the length $\delta g = I \cdot \Delta x$. $k_{st} = 100$, $Q = 200$ l/s. For comparison, the solution of the cubic equation for the frictionless case is drawn

from the now known water depth h. The exit velocities are shown in Fig. 3.12 for different slopes. Since from Fig. 3.11, it is not clear after which height difference δg a certain water depth h is reached, the values of this figure were again converted with $\delta g = \Delta x \cdot I$. These results are shown in Fig. 3.13.

Water depths below $h = 0.125$ m are desirable. This water depth corresponds approximately to the jet width of the exiting, free-falling jet. A jet of 0.125 m is narrow enough to ensure good ventilation of the cells.

Here, the water depths for a flow of $Q = 200$ l/s ($b_0 = 1$ m) are of interest, as this flow is approximately the specific discharge capacity ($Q_s = 0.2 \ldots 0.22$ m^3/(s·m)) of an overshot waterwheel.

The water depth h varies very little with changing gradient I (see Fig. 3.13). Thus, it is simple to achieve desired value of $h = 0.125$ m. A low water depth gives a high exit velocity. Therefore, a value of $\delta g = 0.1 \ldots 0.3$ with a gradient $I < 0.04$ should be used.

One must bear in mind that one wants to keep the head loss due to the channel small. A channel that is too long with a low gradient leads to increased friction and thus to losses. In addition, there are increased costs for the supports and the chute.

However, the channel must not be designed too steeply, otherwise undesired deflections occur. The length of the inclined channel section can be easily calculated with $L_G = \delta g / I$. The application of the equations and figures given above will be shown in the next example for of a channel with a chute.

Example 3.4. ***(Complete dimensioning of a broad-crested weir with subsequent inclined steel chute and a straight outlet.)***

The values from the previous examples 3.2 and 3.3 are retained. Thus, the channel width $b_0 = 1$ m **and the flow** $Q = 0.2$ m^3/s **are kept.**

From the previous example 3.2, the following previously calculated values are taken:

- the overflow height $h_{ov} = 0.252$ m,
- the weir height $w = 0.25$ m,
- the inflow speed $c_0 = 0.4$ m/s,
- the limit depth $h_{gr} = 0.16$ m and
- the limit speed $c_{gr} = 1.253$ m/s.

The length of the weir crown should be $4 \ldots 5 \cdot h_{ov} = 1.0 \ldots 1.26$ m. Now, for the inclined part of the channel, the channel drop δg and the length of the inclined channel L_G are to be determined (see Fig. 3.9). For this, Fig. 3.12 is used. The inclined part of the channel has to be short. The circumferential speed of an overshot waterwheel is about $u_a = 1.5 \ldots 2.2$ m/s. Additionally, the ratio between circumferential speed and exit speed at the end of the chute should be $u_a/c_{2A} = 0.56 \ldots 0.82$ at point A. So a speed of

$$c_{2A} = \frac{u_a}{0.56 \ldots 0.82} = 1.83 \ldots 3.92 \text{ m/s}$$

is required as exit velocity at the end of the channel. For an open channel supply, this range reduces to $c_{2A} = 1.88 \ldots 3.0$ m/s to obtain a short length of the chute L_G. This is the shaded range found in Fig. 3.12. Only gradients of $I = 0.05 \ldots 0.2$ should to be used. If one chooses gradient $I = 0.175$ and exit speed $c_{2A} = 3.0$ m/s, a length of $\Delta x = L_G = 2.9$ m must be provided. The necessary height difference is:

$$\delta g = L_G \cdot I = 2.9 \text{ m} \cdot 0.175 = 0.507 \text{ m}$$

This value is not acceptable (head loss, water supply as high as possible). If the exit speed is reduced to $c_{2A} = 2.0$ m/s, one obtains $L_G = 1.1$ m and $\delta g = L_G \cdot I = 0.1925$ m at $I = 0.175$. The water depth is taken from Fig. 3.11 and is $h_{2A} = 0.1$ m at the end of the inclined channel. The condition $h_{2A} \leq 0.125$ m is thus fulfilled. The channel might be supplemented by a straight outflowing section.

The entire calculated arrangement is displayed in Fig. 3.14. The flow characteristic depends only on the overfall height of the broad crested weir. The characteristic can be calculated using Poleni's formula $Q = C \cdot b_0 \cdot h^{3/2}$, where the flow coefficient C varies with varying overfall height.

One obtains the flow rate values in Table 3.2 by using Fig. 3.7. Since the overflow height h varies with the flow rate, one loses available head difference for small flows. This is head loss.

The arrangement chosen here can also be used as metering of the flow rate. For this purpose a gauge meter is installed in the tail water. It is important that a change in the flow condition from subcritical to critical conditions takes place at the end of the flat weir crown. Under these conditions there is no feedback from the outflowing jet to the approach region.

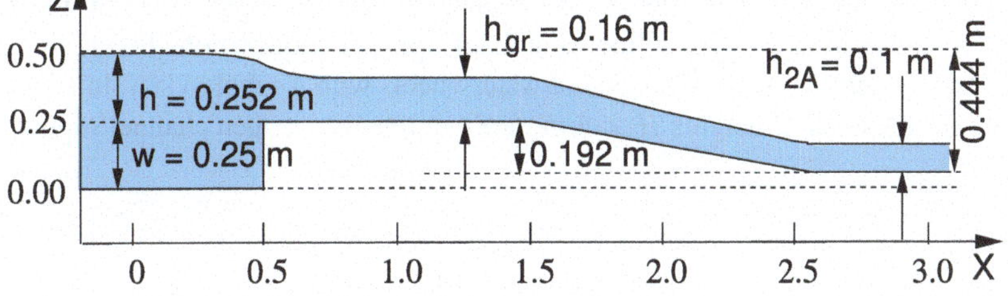

Fig. 3.14 Overall arrangement of the broad-crested weir with a chute

Table 3.2 Discharge values of the arrangement

h [m]	h/w [1]	C [$m^{1/2}/s$]	Q [m^3/s]
0.05	0.2	1.45	0.016
0.10	0.4	1.49	0.047
0.15	0.6	1.52	0.088
0.20	0.8	1.56	0.139
0.25	1.0	1.59	0.199
0.30	1.2	1.62	0.266
0.35	1.4	1.66	0.343

The supply of water via an open channel has the following disadvantages:

1. the water height h and the speed c_1 in the channel change with the flow Q. For smaller flows, a loss of head difference must be accepted. There is also a hydraulic loss $h_{V,bw}$ when water enters the channel. However, this loss can be kept small by appropriate design of the leading edge of the weir crest.
2. the exit speed depends on the flow.

The advantages of channel supply are:

1. simplicity.
2. the arrangement can be used to measure the flow. Thus, accurate compliance with the required design flow can be achieved.
3. a hydraulic recalculation with the software "WinFlume" [3] offers a simple check of one's results.
4. there is no need to regulate the upper water level if the maximum flow can be discharged in the channel. The control of the flow then takes place at the water extraction structure and via the free run, which diverts excess water away from the wheel[8] or takes over the entire water flow Q.
5. the design is very resistant to contamination, and due to the absence of construction elements in the flow path, it is permeable even for larger floating debris.
6. an overshot waterwheel with supply via an open channel can be operated without a rake.
7. the arrangement ensures a safe transport for fish and small organisms downstream, and is therefore fish-friendly.

These advantages outweigh the disadvantages, as nowadays attention must be paid to low costs, fish-friendliness, and unsupervised operation. For high head differences, one should build overshot waterwheels with a supply via a sluice gate,[9] while for lower head heights $H < 4$ m, one can resort to a open channel supply.[10]

[8] In case of wheel standstill, there must be emergency sluices provided.

[9] A constant upper water level, small wheel width can be achieved.

[10] One has a variable upper water level, but small losses relative to the head difference.

3.5 Absolute Path and Intersection Points

3.5.1 Calculation of the Absolute Path

So far, only the water supply to the wheel has been discussed. After the water has left the sluice gate, the tongue inlet or the open channel, the water falls as a free falling jet at a certain angle of attack γ into the waterwheel.

The path of the freely falling water jet will be called the absolute path. The exit point of the centerline of the jet from the respective opening is point A. The entry of the centerline of the jet into the wheel occurs at point B.

The equation of a falling water element of mass $m = \varrho \cdot V$ is given as:

$$\varrho \cdot V \cdot \frac{d^2\vec{r}}{dt^2} = \varrho \cdot V \cdot g \cdot (-\vec{k}) = \vec{F_G}$$

$$\frac{d^2\vec{r}}{dt^2} = g \cdot (-\vec{k})$$

We assume constant pressure inside the jet. The internal pressure distribution and the internal friction in the jet are neglected.

The weight force acts in the direction of the negative Z-axis $-\vec{k}$. By integrating the equation twice with respect to time, one obtains the absolute path of the jet. The boundary conditions are $t = 0$ at point A:[11]

$$x\,(t = 0) = x_A = \delta x$$

$$z\,(t = 0) = z_A = R_a + \delta z + m + \frac{s_0}{2}$$

$$c_{2A,x} = c_{2A} \cdot \cos\,\gamma$$

$$c_{2A,z} = c_{2A} \cdot \sin\,\gamma$$

γ is the angle that the metal sheet of the opening encloses with the X-axis. For $\gamma = 0°$ this is a side opening (sluice gate) and for $\gamma = -90°$, a bottom opening is present.

s_0 is the jet width at point A, m the thickness of the inlet metal plate and δz the distance to the upper vertex of the waterwheel. δx is the horizontal distance between the upper vertex of the wheel (point S) and the exit point A.

The equations for the path of the water jet obtained by integration are:

$$x(t) = c_{2A} \cdot \cos\,(\gamma) \cdot t + x_A \tag{3.11}$$

$$z(t) = -\frac{g}{2} \cdot t^2 + c_{2A} \cdot \sin\,(\gamma) \cdot t + z_A \tag{3.12}$$

[11] The origin of the coordinate system is at point M, the center of the circle of the waterwheel.

where x_A and z_A are the coordinates of the exit point A. The velocities at any point along the path are:

$$c_{2,x}(t) = c_{2A} \cdot \cos \gamma \tag{3.13}$$

$$c_{2,z}(t) = -g \cdot t + c_{2A} \cdot \sin \gamma \tag{3.14}$$

For each time t, the location and velocity of the water element can now be determined. The time t can be eliminated from the equations by rearranging $x(t)$ for t:

$$t = \frac{x - x_A}{c_{2A} \cdot \cos \gamma} \tag{3.15}$$

Substituting this result into $z(t)$ yields:

$$z(x) = -\frac{g}{2} \cdot \left(\frac{x - x_A}{c_{2A} \cdot \cos \gamma} \right)^2 + (x - x_A) \cdot \tan \gamma + z_A \tag{3.16}$$

The obtained solution is a parabola and applies to the center of the water jet. The upper and lower jet boundaries are obtained by adding the value $\pm s_0/2$ to z_A. In Fig. 3.15, the absolute path of the water jet is plotted for various exit angles and exit openings.

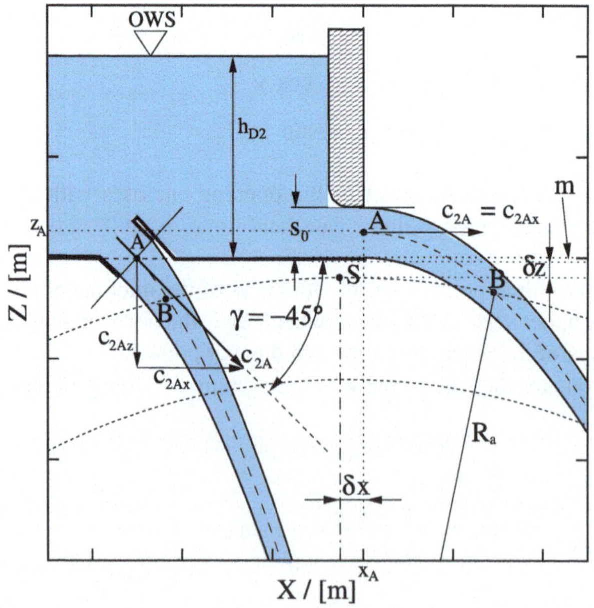

Fig. 3.15 Calculated path of the water jet after exiting a side opening with $\gamma = 0°$ or $\gamma = 45°$

The results for the flow rate Q (jet width s_0) and the entry speed c_{2A} from the previous sections can be used to calculate the shape of the water jet.

3.5.2 The Intersection Points B, C

The intersection points of the centerline of the water jet are plotted in Fig. 3.16. These points are of interest for the design of the waterwheel. Point A is the exit point from the channel. The coordinates $A(x_A, z_A)$ are given.

Point B is the entry point into the waterwheel. This point should be as close as possible to the upper vertex S or exactly at the coordinates of S. This position ensures a good start-up behavior of the waterwheel.

This is especially important in unsupervised operation, as often required today. Bach [4] recommends placing point B even slightly before point S to increase the efficiency a bit more. However, a start-up aid must be provided to set the waterwheel in the correct direction of rotation, which precludes unsupervised operation.

Losses at the entrance will further considered when the position of point B is determined. The points B_1 and B_2 determine the width of the inlet arc b_S and the

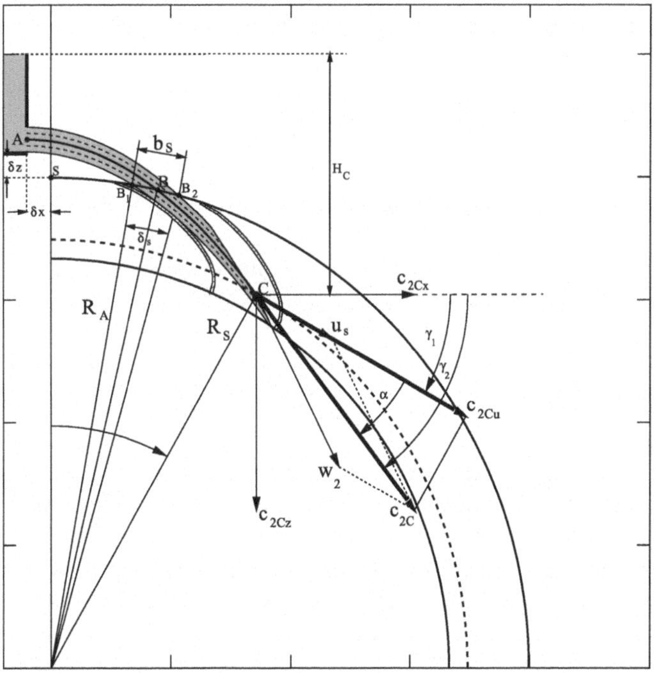

Fig. 3.16 Intersection points B, B1, B2 and C with the center line of the water jet after the exit from a sluice gate

angle δ_S. The inlet arc is calculated from the coordinates of the points $B_1(x_{B1}, z_{B1})$, $B_2(x_{B2}, z_{B2})$ as follows:

$$b_S = \frac{2 \cdot \pi \cdot Ra}{360^\circ} \cdot \delta_S$$

$$\delta_S = \arccos\left(\frac{x_{B2}}{R_a}\right) - \arccos\left(\frac{x_{B1}}{R_a}\right)$$

The width of the inlet arc b_S determines the distance of the cells, i.e. the cell division t_z and thus also the number of cells z and the mutual overlap of the cells. The inlet arc b_S must be smaller than the cell division t_z. Otherwise, the ventilation of the jet is insufficient, as will be shown later.

All previously described intersection points B_1, B_2 and B have in common that they result from the intersection of a circle with a parabola. The equation for an arbitrary circle with radius R and the center M at the origin of the coordinate system is:

$$R^2 = x^2 + z^2$$

The equation for the parabola is given by Eq. (3.16). If one inserts both equations into each other and eliminates the z-coordinate, one obtains a 4th order polynomial, which is cumbersome to solve analytically. The resulting equation is:

$$F(x) = 0 = x^2 - R^2 + \left(-\frac{g}{2} \cdot \left(\frac{x - x_A}{c_{2A} \cdot \cos\gamma}\right)^2 - (x - x_A) \cdot \tan\gamma + z_A\right)^2$$

For this reason, the equations are solved numerically using Newton's method:

$$x_0 = x_S + 0.1\ \text{m} \quad \text{as initial value}$$

$$x_{i+1} = x_i - \frac{F(x_i)}{F'(x_i)}$$

One can iterate until an error limit, e.g. $|x_{i+1} - x_i| < 10^{-8}$, is undercut, obtaining the final value x_e.

Newton's method converges very quickly for the given problem. However, it should be noted that the equation can have a maximum of four roots[12] and thus four intersection points can be calculated. The desired solution can be determined, however, by choosing suitable starting values for the procedure. The final value x_e is obtained via Eq. (3.16). This value corresponds to the function value z.

[12] The following results are possible: (a) four roots if the parabola cuts the circle four times, (b) two roots if the vertex of the parabola lies inside the circle or touches the circle twice, (c) three roots if the circle is touched once and cut twice, (d) one root if the vertex of the parabola touches the circle and (e) no root if the circle and the parabola never touch.

The calculation method is applied for all points, i.e. $B(x_B, z_B)$, $B_1(x_{B1}, z_{B1})$, $B_2(x_{B2}, z_{B2})$, $C(x_C, z_C)$.

The velocities at the respective points can be determined via Eqs. (3.13) and (3.14), if the time is calculated with Eq. (3.15).

Of particular interest are the velocities at point C (intersection of the center line of the water jet with the center of gravity circle), as the losses at the inlet are determined at this point. The coordinates and velocities at point C can be determined with the method presented above.

Example 3.5. ***(Determination of the Velocity at Point*** C***)*** **The water is fed via a sluice gate. The maximum flow rate is** $Q = 0.2\ \frac{\text{m}^3}{\text{s}}$ **and the head difference is** $H = 4.5$ m.

The following information is given about the waterwheel. The wheel radius is $R_a = 2.0$ m, **the width is** $B = 0.90$ m **and rim width is given by** $a = 0.33$ m. **The filling factor of the cells is assumed to be** $\varepsilon = 0.4$ **and rotational speed** $n = 8\ \text{min}^{-1}$. **The upstream water level in the channel is** $h_{D2} = 0.4$ m **above the bottom of the cute. The thickness of the inlet plate is** $m = 0.005$ m, **distance between the upper vertex** S **and the inlet plate in** Z**-direction is** $\delta z = 0.1$ m, **and the distance between the upper vertex and the inlet plate in** X**-direction is** $\delta x = -0.1$ m.

The following values have to be determined: the velocities c_{2C}, $c_{2C,x}$, $c_{2C,z}$, **the projection of the velocity component onto the circumferential direction** $c_{2C,u}$, **the circumferential velocity at the center of gravity** u_s **and the relative velocity** w_{2C} **at point** C **(see Fig. 3.16, all information is drawn to scale in this drawing).**

At the beginning, the exit velocity at point A has to be determined (the dimensionless velocity coefficient is assumed to be $\varphi = 0.95$):

$$c_{2A} = \varphi \cdot \sqrt{2 \cdot g \cdot h_{D2}} = 0.95 \cdot \sqrt{2 \cdot 9.81 \cdot 0.4} = 2.661\ \text{m/s}$$

$$s_0 = Q/(b_0 \cdot c_{2A}) = 0.084\ \text{m}$$

The coordinates of point A are $A(x_A = -0.1, z_A = 0.1 + R_a + s_0/2 = 2.142\ \text{m})$. Then the radius of the centroid circle is calculated to be:

$$R_s = \sqrt{\frac{Q \cdot 60}{2 \cdot \pi \cdot n \cdot B} + (R_a - a)^2} = \sqrt{\frac{0.2 \cdot 60}{2 \cdot \pi \cdot 8 \cdot 0.9} + (2 - 0.33)^2} = 1.748\ \text{m}$$

With this value, the x_C coordinate of the intersection point C can now be determined using Newton's method:

$$x_0 = 0.2\ \text{m} \quad \text{initial value}$$

$$F(x_i) = R_s^2 - x_i^2 - \left(-\frac{g}{2} \cdot \left(\frac{x_i - x_A}{c_{2A} \cdot \cos\gamma}\right)^2 + (x_i - x_A) \cdot \tan\gamma + z_A\right)^2$$

$$F'(x_i) = -2 \cdot x_i - 2 \cdot \left(-\frac{g}{2} \cdot \left(\frac{x_i - x_A}{c_{2A} \cdot \cos\gamma}\right)^2 + (x_i - x_A) \cdot \tan\gamma + z_A\right) \cdot \left(-g \cdot (x_i - x_A)/(c_{2A}^2 \cdot \cos^2\gamma) + \tan\gamma\right)$$

$$x_{i+1} = x_i - \frac{F(x_i)}{F'(x_i)}$$

The x_C coordinate of the intersection point C is $x_C = 0.873$ m. The Z-coordinate is obtained via the equation:

$$z_C = -\frac{g}{2} \cdot \left(\frac{x_C - x_A}{c_{2A} \cdot \cos \gamma}\right)^2 + \tan \gamma \cdot (x_C - x_A) + z_A$$
$$= -\frac{9.81}{2} \cdot \left(\frac{0.973}{2.661}\right)^2 + 2.142 = 1.486 \text{ m}$$

The time it takes for a water particle to travel from point A to point C is determined with the coordinates x_C and x_A:

$$t = \frac{x_C - x_A}{c_{2A} \cdot \cos \gamma} = \frac{0.873 + 0.1}{2.661} = 0.366 \text{ s}$$

Now, the various velocity components can be specified:

$$c_{2C,x} = c_{2A} \cdot \cos \gamma = 2.661 \text{ m/s}$$
$$c_{2C,z} = -g \cdot t + c_{2A} \cdot \sin \gamma = -9.81 \cdot 0.366 = -3.59 \text{ m/s}$$
$$c_{2C} = \sqrt{c_{2Cx}^2 + c_{2Cz}^2} = \sqrt{2.661^2 + 3.59^2} = 4.469 \text{ m/s}$$

The angle α between the vector $\vec{u}_s$ and $\vec{c}_{2C}$ is needed to determine the circumferential component of c_{2C}.

This angle results from:

$$\gamma_1 = 90° - \arctan\left(\frac{z_C}{x_C}\right) = 90° - \arctan\left(\frac{1.486}{0.873}\right) = 30.43°$$
$$\gamma_2 = -\arctan\left(\frac{c_{2Cz}}{c_{2Cx}}\right) = -\arctan\left(\frac{-3.59}{2.661}\right) = 53.45°$$
$$\alpha = \gamma_2 - \gamma_1 = 53.45° - 30.43° = 23.02°$$

Now, the component of c_{2C} can be determined, which is projected onto the circumference:

$$c_{2C,u} = c_{2C} \cdot \cos \alpha = 4.469 \cdot \cos 23.02° = 4.113 \text{ m/s}$$

This velocity component drives the wheel during the startup operation. Later, the resting water fillings in the cells additionally drive the wheel. The circumferential speed at the center of gravity is given by:

$$u_s = R_s \cdot \omega = R_s \cdot \frac{2 \cdot \pi \cdot n}{60} = 1.748 \cdot \frac{2 \cdot \pi \cdot 8}{60} = 1.464 \text{ m/s}$$

The relative speed w_2 can now be determined by applying the cosine theorem:

$$w_{2C} = \sqrt{u_s^2 + c_{2C}^2 - 2 \cdot c_{2C} \cdot u_s \cdot \cos \alpha}$$
$$= \sqrt{1.464^2 + 4.469^2 - 2 \cdot 4.469 \cdot 1.464 \cdot \cos 23.02°} = 3.174 \text{ m/s}$$

All speeds of interest at point C have been calculated with the equations above. The velocity vectors are shown in Fig. 3.16. Care should be taken in the design to keep $c_{2C,u}$ as small as possible. This can be done by decreasing the vertical distance between the points A and C.

3.6 Relative Path, Cell Ventilation and Noise

Figure 3.17 shows the passage of a cell through the water jet at different but equal time steps. The area extending beyond the entry arc between points B and B_1, B_2

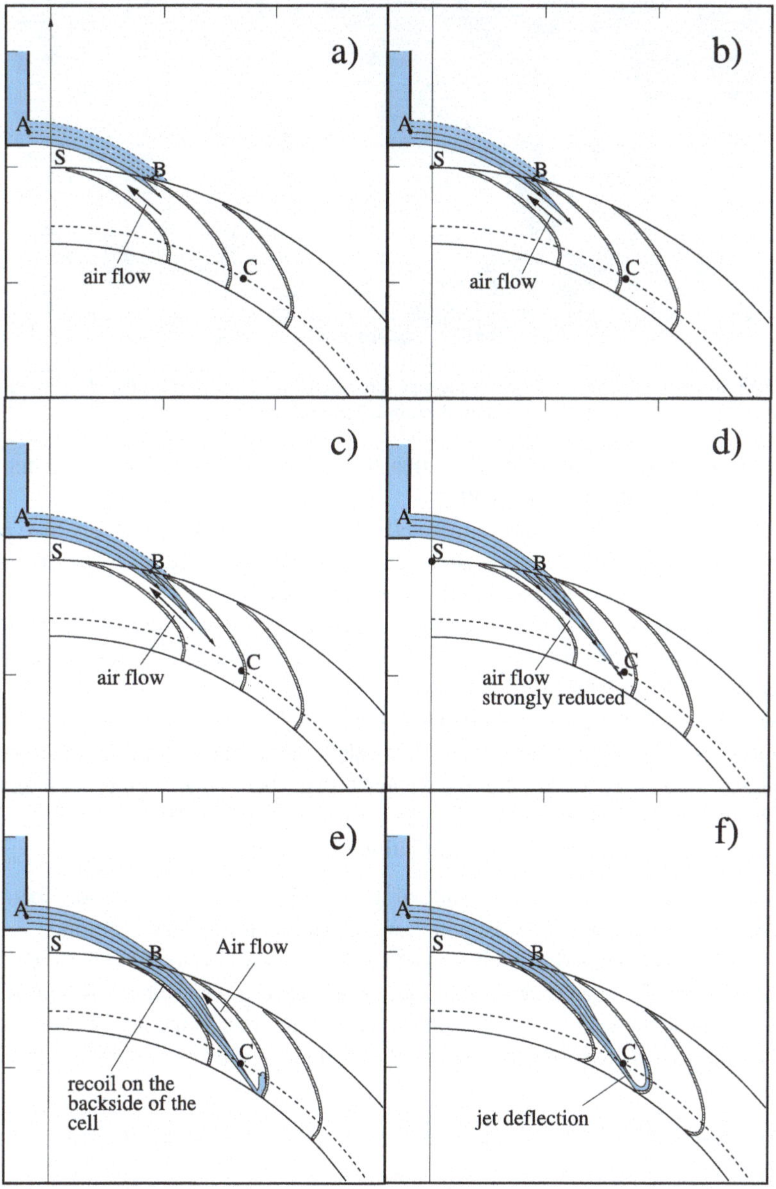

Fig. 3.17 Movement of a cell through the water jet

Fig. 3.18 Jet behavior during the cell's passage through the jet: Note the recoil and the ventilation of the cell at the bottom of the jet (waterwheel mill Stedten/Thuringia)

corresponds to the volume of water that has entered the cell. The water that has entered displaces the air in the cell.

The illustration clearly shows that cell ventilation cannot exclusively take place over the entry arc. First, the upper area of the cell space is closed off by the incoming water jet. The air is compressed in this part of the cell.

The air exits below the jet until the volume is completely closed. In the lower part, this happens over the entry arc between the cell back and the underside of the water jet (see Figs. 3.18 and 3.19).

Afterwards, in the lower area between the blade back and the water jet, the air is compressed. The compression of the air leads to break through of the enclosed air through the water jet, so that the water entry is disturbed. A wide entry jet manifests itself in a thumping noise.

There are two possibilities for ventilation of the cells:

1. One can open the cells to the inside of the wheel and create ventilation openings. This leads to a hydraulic loss, which makes this an unfavorable option.
2. The wheel is designed in such a way that the incoming water jet is narrower than the wheel and air can escape at the open side surfaces. This causes difficulties for wide wheels which are usually split into different compartments. One solution is that the wheels are subjected to several partial jets, to ensure sufficient ventilation to the sides.

According to Bach [4], for the above reasons, the wheel width B must be chosen to be:

$$B = b_0 + 0.1 \ldots 0.2\ \mathrm{m} \tag{3.17}$$

Where b_0 is the width of the water jet.

Fig. 3.19 Jet behavior during the cell's passage through the jet: Ventilation of the cell at the top of the jet (waterwheel mill Stedten/Thuringia)

In Fig. 3.17 the relative direction of movement of the water can also be seen (marked by arrows).

While so far, the absolute movement relative to the stationary space was calculated (parabola of the water jet), the relative path of the water should also be determined.

The relative path results from the movement of the water in relation to the rotating waterwheel. The relative speed is then the speed which the water has in relation to the rotating waterwheel.

The relative speed w_2 results from vector addition of the respective peripheral speed $u = \omega \cdot R$ and the inlet speed c_2 at any points of entry (e.g. at point B or C, see Fig. 3.16).

The parallelogram of the velocity components is drawn for point C. An observer rotating with the wheel would not perceive the parabolic shape of the water jet, but the direction of movement of the water marked by arrows in Fig. 3.17.

The relative path curves are determined by finding the intersection points of the various absolute water paths with the outer circle (radius R_a). Let t_0 be the time that elapses between water exit to the entry point of the water jet into the wheel. The time elapsed between B_1 and B_2 (see Fig. 3.16) is given by:

$$t_1 = \frac{\widehat{B_1 B_2}}{u_a}$$

If this time $t_1 + t_0$ is inserted into the parabola equation of the water jet (3.11) and (3.12), one obtains the location to which the water jet has extended from the entry point into the wheel during the time t_1.

The connecting line between the calculated point $P(x, z)$ and the point B_2 is now a curve of the relative path. This procedure is continued for the following points B_3 and so forth. The relative path curves shown in Fig. 3.17 are obtained.

This method provides the intersection points for the different absolute paths of the water particles with the waterwheel outer circle. It allows us to determine the times t_1, t_2, t_3 etc. After the process of water entry shown in Fig. 3.17 is almost completed, the back of the blade of the next cell hits the water jet and there is a recoil on the following cell. This recoil is opposite to the direction of rotation of the wheel and therefore associated with a loss.

The most favorable cell shape for entry has the form of the relative path of the water. There is no recoil on the blade back for this cell shape. However, the cell curve curved according to the relative path is unfavorable for the exit loss, as the water exits the wheel too early.

Bach [4] therefore recommends choosing the cell shape in such a way that only half of the water that has entered strikes the back of the blade. The cell division should therefore also to be chosen wider than the entry arc length b_S.

The ventilation of the cells is tightly correlated with noise generation of a waterwheel. Waterwheels generally generate noise during water entry, when air is compressed by the water filling the cells. The noise of the overshot waterwheel can be reduced by:

1. water supply by an open channel that reduces the entrance velocity (refer to Fig. 3.20)
2. a sine wave- or V-shaped metal sheet at the entry as shown in Fig. 3.21. The water jet lets the air a path to escape (white colored part of the entry jet) because the jet trajectories vary over the entry.

Fig. 3.20 Open channel water supply with low inflow velocity. Steinfurthsmühle, Königsbrück Sachsen, Germany. With kind permission of Tobias Gappisch, Wasserkraft- und Anlagentechnik, Göda

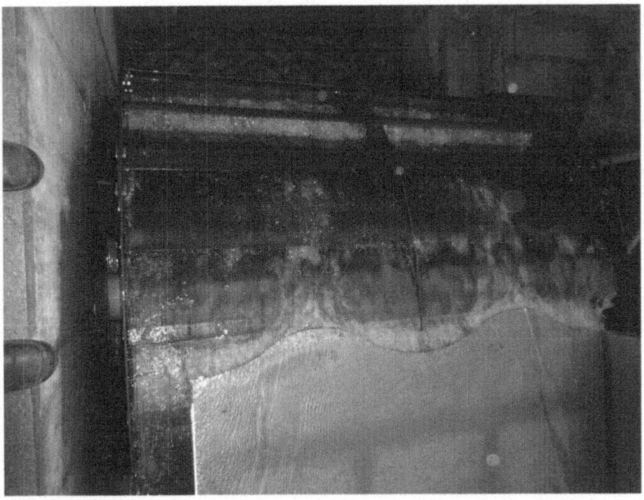

Fig. 3.21 Noise reduction with a curved metal sheet at the water entry. Steinfurthsmühle, Königsbrück Sachsen, Germany. With kind permission of Tobias Gappisch, Wasserkraft- und Anlagentechnik, Göda

Fig. 3.22 Quiet overshot waterwheel running at 80% of the rated output power. Steinfurthsmühle, Königsbrück Saxony, Germany. With kind permission of Tobias Gappisch Wasserkraft- und Anlagentechnik, Göda

Figure 3.22 shows the same waterwheel of Fig. 3.21 at 80% of the rated power output. The noise level is significantly reduced and the water smoothly approaches the wheel compared to Fig. 4.5.

References

1. Bollrich G, Preißler G (1992) Technische Hydromechanik Band 1, 3rd edn. Verlag für Bauwesen, Berlin
2. Aigner D, Horlacher HB (1997) Optimum design of a self-regulated spring steel throtteles for sewer overflow tanks. In: XXVII IAHR congress, San Francisco 1997, pp 75–80
3. Wahl TL, Clemmens AJ, Bos MG, Replogle JA (2013) Winflume, U.S. Department of the Interior, Bureau of Reclamation. www.usbr.gov/pmts/hydraulics_lab/winflume/
4. Bach Cv (1886) Die Wasserräder, 1st edn. Verlag von Konrad Wittwer, Stuttgart

Chapter 4
Discharge Capacity

4.1 Circumferential and Rotational Speed

The peripheral speed at the outer radius of the wheel has to be chosen according to Bach [1]:

$$u_a = 1.5 \ldots 2.2 \text{ m/s} \tag{4.1}$$

These values were confirmed by experiments Meerwarth [2] (refer to Chap. 10, Fig. 10.9).

A study by the author [3] has shown that the optimal peripheral speed can be described with the equation:[1]

$$u_{a,opt} = \begin{cases} 1.23 \cdot \sqrt{R_a} & \text{in } \left[\frac{\text{m}^{1/2}}{\text{s}}\right] \quad \text{for} \quad R_a \leq 3.2 \text{ m} \\ 2.2 \, \frac{\text{m}}{\text{s}} & \text{if} \quad R_a > 3.2 \text{ m} \end{cases} \tag{4.2}$$

The equation was derived by using Froude's similarity law. The rotational speed is then calculated by the equation:

$$n = \frac{u_a \cdot 60}{2 \cdot \pi \cdot R_a} \tag{4.3}$$

Both equations above give a rotational speed between $n = 11.7 \ldots 5.25 \text{ min}^{-1}$ for typical radii of the wheel ranging from $R_a = 1 \ldots 4$ m.

The upper limit for the rotational speed (runaway speed) is three times of the best rotational speed according to Meerwarth [2]. Meerwarth determined the value to be $n = 27 \text{ min}^{-1}$ for a wheel with $D_a = 3.6$ m.

[1] See Chap. 9, section "Scaling rules for waterwheels".

D. M. Nuernbergk, *Overshot Waterwheels*, Springer Tracts in Mechanical Engineering, https://doi.org/10.1007/978-3-031-95385-9_4

If we assume that the potential power is 80% transferred in kinetic power (friction in the bearings, the gear box neglected) $\frac{\dot{m}}{2} \cdot u_a = \dot{m} \cdot g \cdot h$ we can write:

$$n_{runaway} = 0.8 \cdot \frac{60 \cdot \sqrt{g \cdot H}}{2 \cdot \pi \cdot R_a} \tag{4.4}$$

The flow of mass is defined as $\dot{m} = \varrho \cdot \frac{dV}{dt} = \varrho \cdot Q$, where g is the gravitational acceleration and ϱ the density of water.

In this formula, it was assumed that only half of the head difference is active because the water exits the wheel at about the height of the wheel shaft. The centrifugal force is so high that the cells are emptied very early. This behavior is shown in Fig. 4.1 in the lower right corner for the wheel investigated by Meerwarth [2]. The resulting runaway speed is in good agreement with experimental results found by Meerwarth [2] and Quaranta [4].

The forces are drawn to scale in the figure. For the centrifugal acceleration (proportional to the centrifugal force) a value of e.g. $a_z = R_a \cdot \omega^2 = R_a \cdot \left(\frac{2 \cdot \pi \cdot n}{60}\right)^2 = R_a \cdot \left(\frac{2 \cdot \pi \cdot 27}{60}\right)^2 = 14.4\ \text{m/s}^2$ is large compared to the gravitational acceleration

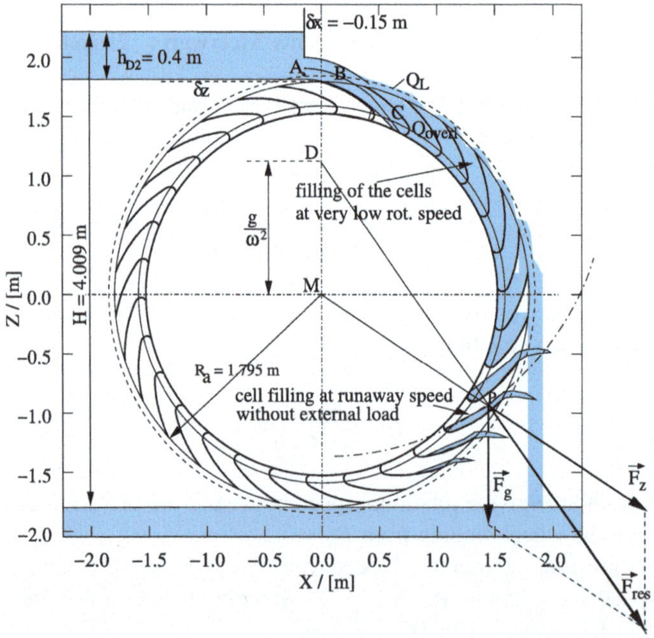

Fig. 4.1 Overload conditions of a waterwheel. The overload of the cell filling due to too low rotational speed is shown in the upper right part of the wheel. There is a leakage flow when the cells are completely filled. The water flows above the filled cells in this case. The cell fillings are shown in the lower right part of the wheel under runaway speed conditions

$g = 9.81\ \text{m/s}^2$. The water level in the cell arranges as a circular surface in the cells because the water particles adjust to the total force which results from the vector sum of the gravitational and the centrifugal force.

The distance between the points D and M is $DM = g/\left(\frac{2\cdot\pi\cdot 27}{60}\right)^2 = 1.23$ m under runaway conditions compared to $DM = 11$ m at optimum rotational speed of $n = 9\ \text{min}^{-1}$. Under optimum rotational speed, the surfaces are almost flat compared to the horizon. Runaway speed is an overload condition for an overshot waterwheel with an efficiency equal to zero. Later it will be shown that there are other overload conditions as well.

4.2 Cell Division and Number of Arms

To determine the shape of the cells the number and division of the cells have to be known. First, the width a of the side rings is required. According to Bach [1], we have:

$$a = \frac{1}{6} \cdot (H)^{1/3} \ldots \frac{1}{4} \cdot (H)^{1/3} \tag{4.5}$$

Common values are $a = 0.25 \ldots 0.4$ m. The cell division is determined according to Henne [5] as follows:

$$t_{z,a} = 0.75 \cdot a + 0.1 \tag{4.6}$$

or according to Bach [1]

$$t_{z,a} = \frac{4}{3} \cdot b_S \ldots \frac{3}{2} \cdot b_S \tag{4.7}$$

The index indicates that it is a preliminary assumption. The number of cells is calculated approximately with the following equations:

$$z_a = \frac{2 \cdot \pi \cdot R_a}{t_{z,a}} \qquad z = \lceil z_a \rceil \tag{4.8}$$

The number of cells and the number of arms should not to be determined independently of each other. With the rounded value, one searches in Table 4.1 and determines the final value of the number of cells z. Once the value for z and the number of arms i have been determined, the division at the outer radius must be recalculated. The final result is:

$$t_z = \frac{2 \cdot \pi \cdot R_a}{z} \tag{4.9}$$

Table 4.1 Cell numbers and number of arms

Condition	Cell number z	Arm number i	Division angle γ_t
$z < 18$	$z = 18$	$i = 6$	$\gamma_t = 20°$
$18 < z < 21$	$z = 18$	$i = 6$	$\gamma_t = 20°$
$21 \leq z < 23$	$z = 24$	$i = 6$	$\gamma_t = 15°$
$24 < z < 26$	$z = 24$	$i = 6$	$\gamma_t = 15°$
$26 \leq z < 30$	$z = 30$	$i = 6$	$\gamma_t = 12°$
$30 < z < 33$	$z = 30$	$i = 6$	$\gamma_t = 12°$
$33 \leq z < 36$	$z = 36$	$i = 6$	$\gamma_t = 10°$
$36 < z < 39$	$z = 36$	$i = 6$	$\gamma_t = 10°$
$39 \leq z < 42$	$z = 45$	$i = 6$	$\gamma_t = 8°$
$42 < z < 46$	$z = 45$	$i = 6$	$\gamma_t = 8°$
$46 \leq z < 48$	$z = 48$	$i = 8$	$\gamma_t = 7.5°$
$48 < z < 52$	$z = 48$	$i = 8$	$\gamma_t = 7.5°$
$52 \leq z < 56$	$z = 56$	$i = 8$	$\gamma_t = 6.43°$
$56 < z < 60$	$z = 56$	$i = 8$	$\gamma_t = 6.43°$
$60 \leq z < 64$	$z = 60$	$i = 8$	$\gamma_t = 6°$
$64 < z < 67$	$z = 60$	$i = 8$	$\gamma_t = 6°$
$67 \leq z < 70$	$z = 72$	$i = 10$	$\gamma_t = 5°$
$70 < z < 75$	$z = 72$	$i = 10$	$\gamma_t = 5°$
$75 \leq z < 80$	$z = 80$	$i = 10$	$\gamma_t = 4.5°$

Furthermore, one strives to obtain integer values for the division angle

$$\gamma_t = \frac{360°}{z}$$

This is guaranteed if the values from Table 4.1 are used.

4.3 Cell Shape

The shape of the cell should meet two requirements:

1. The backlash against the cell blade back should be kept as small as possible.
2. The incoming jet should be completely deflected at the bottom of the cell.
3. The water should exit the cells as close to the lower water as possible.

All demands cannot be completely fulfilled at the same time. According to Henne, the cell construction which is shown in Fig. 4.2, gives good results.

For cells made from sheet metal, the following apply:

$$\widehat{J_1E} = 1.25\ldots1.3 \cdot t_z \qquad \overline{JF} = 0.33\ldots0.4 \cdot a \tag{4.10}$$

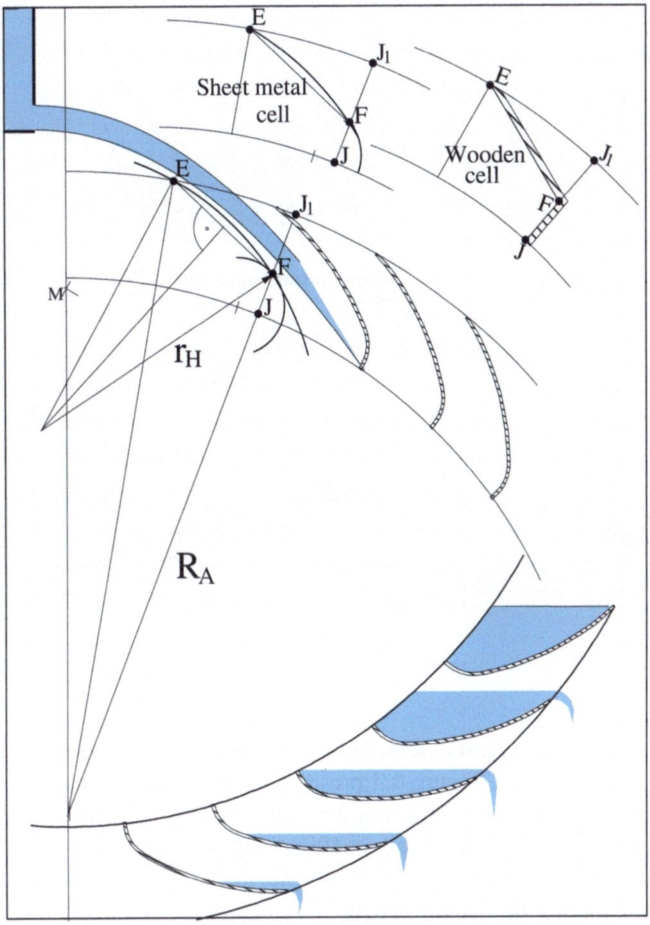

Fig. 4.2 Construction of the blade curve for the cells according to Henne [5]

$$R_H = 2\ldots3 \cdot \overline{FE} = 2\ldots3 \cdot \sqrt{(\widehat{J_1E})^2 + (a - \overline{JF})^2}$$

The straight piece JF is often replaced by a semicircular arc. This measure almost completely deflects the beam (see Fig. 3.17 sub-picture f). Bach [1] uses a quarter circle (Fig. 4.2). For wooden cells, the following applies:

$$\widehat{J_1E} = 1.25 \cdot t_z \qquad \overline{JF} = 0.4\ldots0.5 \cdot a \tag{4.11}$$

The straight piece $\overline{JF}$ is called a bar blade and $\overline{FE}$ a thrust blade. The shape of the cell at point E (entry and exit point of a wooden cell) should be chosen as in Fig. 4.2, as this shape counteracts premature water exit. This is because the water exit over sharp-edge metal sheet is reduced compared to a rounded edge.

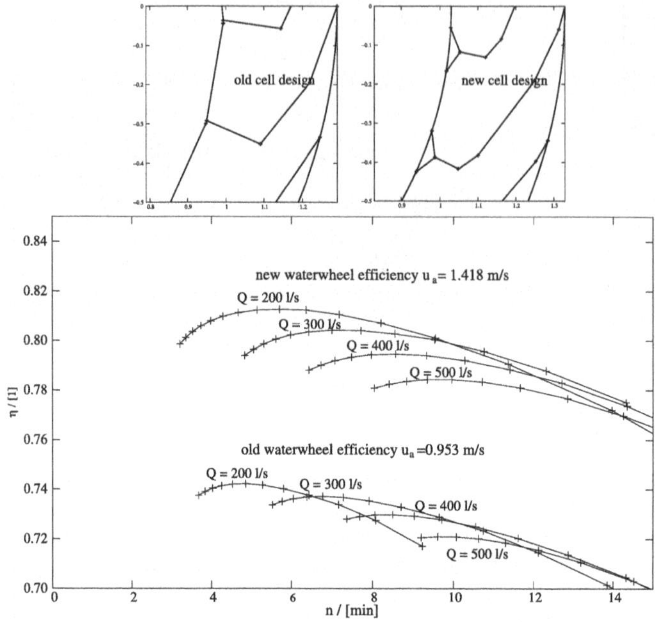

Fig. 4.3 Comparison of the efficiencies of different cell shapes at a particular site

Both cell shapes, i.e., wood and metal cells, are compared if we calculate the exit losses in more detail. Figure 4.3 presents the differences between cell shapes for two wheels which were realized at the same site (the old wheel was replaced), where differences in efficiency of about 10% could be observed.

Wood cells and wooden wheels should no longer be built due to their poor efficiency and a lower discharge capacity.

The poor efficiency is due to the lack of curvature of the thrust and bar blade. This results in a lack of jet deflection. In addition, the exit losses are increased because of the missing curvatures.

The reduced discharge capacity stems from the issue that the material thickness of the wooden blades (15 ... 25 mm) must be made wider than the sheets of the metal cells (2 ... 4 mm) by comparison. In addition, wooden wheels lose their concentricity after a few years in operation.

A compromise between the fully rounded metal and wooden cells are bent sheet metal cells like in Fig. 4.3 (refer also to Fig. 8.10). The bending follow the shape of a rounded metal cell. The manufacturing is technologically simple and allows for easy assembly.

4.4 Specific Discharge- and Discharge-Capacity

To compare the difference in discharge capacity between metal cells and wooden cells, it is easy to calculate the volume of the ring and subtract the volume displaced by the blades.

The volume of a cell can be derived if this volume is divided by the number of cells. The cell volume therefore is given by:

$$V_z = \frac{B \cdot \pi}{z} \cdot (R_a^2 - (R_a - a)^2) \tag{4.12}$$

The thickness of the material is neglected for simplicity.

The water volume which displaces the air in the cells can be calculated as follows:

$$V_q = \frac{Q \cdot 60}{z \cdot n} \tag{4.13}$$

The ratio of the displaced and the cell volume is the filling factor, which is defined as:

$$\varepsilon = \frac{V_q}{V_z} = \frac{Q \cdot 60}{B \cdot \pi \cdot n \cdot (R_a^2 - (R_a - a)^2)} \tag{4.14}$$

Interestingly, the cell filling decreases hyperbolically with the rotational speed and increases linearly with the flow rate. The filling is proportional to the torque of the wheel. Thus, the torque decreases with a lower filling factor or higher rotational speed for a given flow rate.

The maximum flow which the waterwheel can discharge is given in older publications (e.g., Bach [1], Henne [5], Meerwarth [2]):

$$Q_{max} = \varepsilon_a \cdot u_a \cdot B \cdot a \tag{4.15}$$

Thus, the two filling factors ε_a, ε are different by definition. These old publications state that the filling factor should be in the range $\varepsilon_a = 0.25 \ldots 0.5$.

It is necessary to convert the old filling factor to the more useful equation (4.14). The conversion formula is as simple as:

$$\frac{\varepsilon}{\varepsilon_a} = \frac{1}{1 - \frac{a}{2 \cdot R_a}} \tag{4.16}$$

The new fill factors ε in Eq. (4.14) are somewhat larger than the old.

For example, for the parameters $R_a = 2$ m, $a = 0.33$ m, we obtain the filling factor $\varepsilon = 1.09 \cdot \varepsilon_a$. This difference is not negligible. The difference arises because the old equation is based on the outer peripheral speed, while the new equation (4.17) uses the peripheral speed at the centroid.

This leads to the equation for the maximum useful discharge capacity.[2] This is repeated here:

$$Q_{max} = \varepsilon \cdot n \cdot B \cdot \pi \cdot \frac{\left[R_a^2 - (R_a - a)^2 - \frac{(\widehat{EF}+\widehat{FJ})\cdot b_{Mat}\cdot z}{\pi}\right]}{60} \tag{4.17}$$

The advantage of Eq. (4.17) over (4.15) is that the volume reduction of the cell due to the material thickness of the cell walls is taken into account.

If one uses the old fill factors $\varepsilon_a = 0.25 \ldots 0.5$ in Eq. (4.17), one gets fewer filled cells, which can be desirable concerning exit loss. But one has to consider that wheel width increases which increase the total costs of the wheel.

Furthermore, the specific discharge capacity per unit meter can be determined using Eqs. (4.17) and (4.2). The wheel width B is set to 1 m and the rotational speed is replaced according (4.3). The specific discharge capacity is then given by:

$$Q_s = \varepsilon \cdot 0.615 \cdot \frac{\left[R_a^2 - (R_a - a)^2 - \frac{(\widehat{EF}+\widehat{FJ})\cdot b_{Mat}\cdot z}{\pi}\right]}{\sqrt{R_a}} \tag{4.18}$$

The specific discharge capacity is shown for different overshot waterwheels depending on the radius and on the width of the ring a in Fig. 4.4. The material thickness of the cells was not taken into account in this figure. Furthermore, it was assumed that a wheel with $R_a \leq 3 \cdot a$ cannot be realized by construction.

A specific discharge capacity of $Q_s = 0.13 \ldots 0.4$ m^3/(s · m) results for typical ring widths of $a = 0.25 \ldots 0.4$ m.

It can be seen that for larger wheel radii, a higher rim width can be allowed in order to achieve a high specific discharge capacity. Larger waterwheels are then narrower and have a smaller construction volume. However, it should be noted that with a larger rim width, the entry loss also increases, as will be shown in the next section.

Furthermore, it can be observed that overshot waterwheels cannot be used for head differences below $H = 2 \ldots 2.5$ m, as only a small ring width a can be allowed if a good efficiency is to be achieved.

[2] A similar expression can be derived for Zuppinger waterwheels as it was shown in [6].

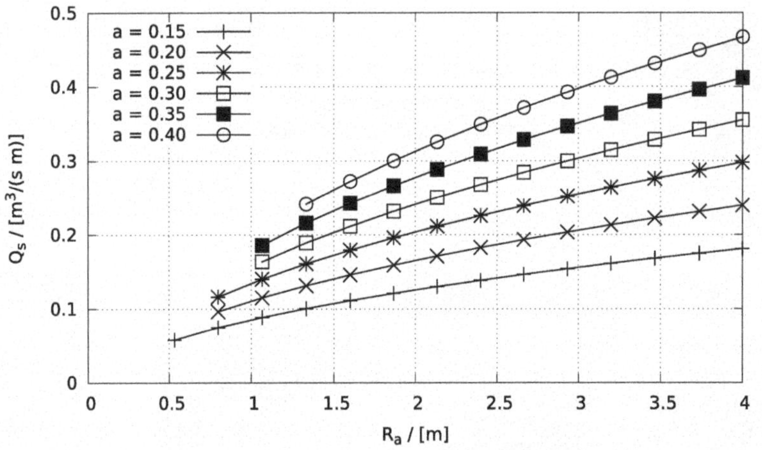

Fig. 4.4 Specific discharge capacity of overshot waterwheels depending on the wheel radius R_a at various ring widths a. It is assumed that the wheel radius cannot be smaller than $R_a = 3 \cdot a$

However, at such locations, the flows to be discharged are usually high. Zuppinger waterwheels [6] or hydropower power screws [7], [8] are used in this particular case instead.

Meerwarth has measured a overload filling factor of $\epsilon_{overl} = 0.72$ (value already converted using Eq. (4.16)) where the waterwheel is completely overloaded. Above this filling factor the water is flowing over the cells without filling the cells and without generation of torque as it is shown in Fig. 4.1 in the upper right quadrant. There is then an additional leakage component Q_L.

The overload flow rate is defined as:

$$Q_{overl} = 0.72 \cdot n \cdot B \cdot \pi \cdot \frac{\left[R_a^2 - (R_a - a)^2\right]}{60} \tag{4.19}$$

This equation was derived in Nuernbergk [9]. Again, the overload flow rate is linearly dependent on the rotational speed. It means that the wheel can be operated with variable rotational speed up to overloading without compromising efficiency.

Figure 10.3 shows the mechanical output power of a waterwheel which was investigated by Meerwarth [2]. The agreement between measurements and calculation results using Eq. (4.19) is very good.

Figure 4.5 shows an overloaded overshot waterwheel. The wheel was running very slowly and could not develop the power output which was estimated. Consequently, the old wheel was dismantled and scrapped. It was later replaced by a new wheel designed according to the rules in this book. The behavior of the new wheel is shown in Figs. 3.20, 3.21, and 3.22.

Fig. 4.5 Slowly running overloaded waterwheel (With kind permission of Tobias Gappisch Wasserkraft- und Anlagentechnik, Göda)

References

1. Bach Cv (1886) Die Wasserräder, 1st edn. Verlag von Konrad Wittwer, Stuttgart
2. Meerwarth KW (1935) Experimentelle und theoretische Untersuchungen am oberschlächtigen Wasserrad. PhD thesis, Technische Hochschule Stuttgart
3. Nuernbergk DM (2009) Modellgesetze für Wasserräder und ihre Anwendung auf das oberschlächtige Wasserrad. Wasserkraft Energie Heft 3:38–46
4. Quaranta E, Revelli R (2015) Output power and power losses estimation for an overshot water wheel. Renew Energy 83:979–987
5. Henne H (1903) Die Wasserräder und Turbinen – ihre Berechnung und Konstruktion. Verlag von Bernh. Friedr. Voigt, Leipzig
6. Nuernbergk DM (2005) Wasserräder mit Kropfgerinne – Berechnungsgrundlagen und neue Erkenntnisse, 1st edn. Verlag Moritz Schäfer, Detmold
7. Nuernbergk DM (2020) Wasserkraftschnecken – Berechnung und Entwurf von archimedischen Schnecken als Wasserkraftmaschine, 2nd edn. Verlag Moritz Schäfer, Detmold
8. Nuernbergk DM (2023) Hydropower screws – calculation and optimal design of Archimedean screws, 1st edn. Springer, Cham
9. Nuernbergk DM, Gappisch T, Pommer M (2023) Das Überlastverhalten von oberschlächtigen Wasserrädern. Wasserwirtschaft 12:12–19

Chapter 5
Hydraulic Losses

5.1 Determination of the Losses at the Entrance

It is important for the calculation of the entry losses to determine the coordinates and the velocities at point C. The intersection of the parabola with the centroid circle has to be determined for this purpose.

It is assumed that the conversion of the jet momentum into a torque occurs at the intersection point. Friction is neglected.

The component of the velocity vector c_{2C}, projected onto the circumference, is the torque-generating component of the incoming water jet. Since the jet is fully deflected by the circular curvature of the blade curve, it applies for the backward directed relative velocity $w_{3C} = c_{2C,u} - u_s$ (see Fig. 5.1). The reaction force acting on the blade at point C in the circumferential direction due to the difference in momentum flows is therefore:

$$\begin{aligned} F_{R,\perp} &= \varrho \cdot Q \cdot \left(c_{2C,u} - u_s\right) - \varrho \cdot Q \cdot \left(w_{3C,u}\right) \cdot \cos\gamma \\ &= \varrho \cdot Q \cdot \left(c_{2C,u} - u_s\right) - \varrho \cdot Q \cdot \left(c_{2C,u} - u_s\right) \cdot \cos\gamma \end{aligned}$$

The torque is now calculated as:

$$M = F_{R,\perp} \cdot R_s$$

This results in the converted power:

$$P = M \cdot \omega = M \cdot \frac{2 \cdot \pi \cdot n}{60} = F_{R,\perp} \cdot R_s \cdot \frac{u_s}{R_s} = F_{R,\perp} \cdot u_s$$

where the relationship $u_s = R_s \cdot \omega$ was used. To obtain an energy height, set:

$$P = F_{R,\perp} \cdot u_s = \varrho \cdot g \cdot Q \cdot h$$

D. M. Nuernbergk, *Overshot Waterwheels*, Springer Tracts in Mechanical Engineering, https://doi.org/10.1007/978-3-031-95385-9_5

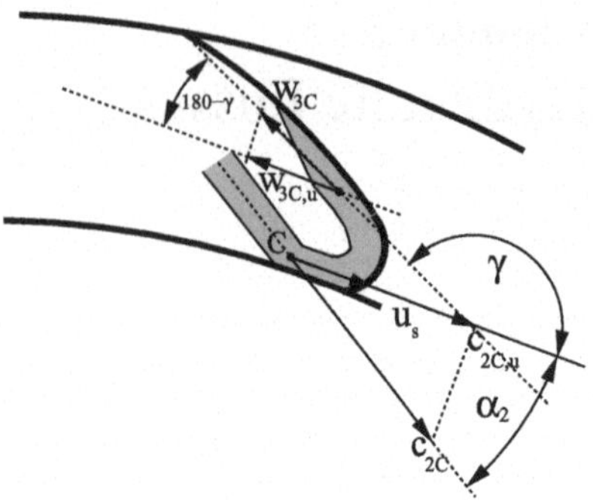

Fig. 5.1 Speeds during the deflection of the water jet in a fully rounded cell

When the value for $F_{R,\perp}$ is substituted, we finally get the value for the converted energy height:

$$h = \frac{F_{R,\perp} \cdot u_s}{\varrho \cdot g \cdot Q} = \frac{\left(c_{2C,u} - u_s\right) \cdot u_s \cdot (1 - \cos\gamma)}{g}$$

This height h of the total height difference $H_C = R_a + \delta z + m + h_{D2} - z_C$ between the upper water level and point C is given to the wheel as a force impulse by the momentum of the water jet (see Fig. 3.16 for the definition of H_C). We can now determine the entry loss $h_{V,e1}$ by forming the difference between the available difference H_C and the value h. We obtain:

$$h_{V,e1} = H_C - h = H_C - \frac{\left(c_{2C,u} - u_s\right) \cdot u_s \cdot (1 - \cos\gamma)}{g}$$

This loss is a combined loss, as all losses from the upper water level to point C are taken into account.[1] For a wooden cell $\gamma = 90°$, the equation simplifies to:

$$h_{V,e1} = H_C - \frac{\left(c_{2C,u} - u_s\right) \cdot u_s}{g} \qquad \text{loss: perpendicular wooden cell} \qquad (5.1)$$

since the water jet hits the vertically mounted impact blade.[2] For a rounded sheet metal cell, which almost realizes a deflection of the water jet of $\gamma = 180°$, one simply writes (assuming $c_{2C,u} = c_{2C,uu}$):

$$h_{V,e1} = H_C - K \cdot \frac{\left(c_{2C,u} - u_s\right) \cdot u_s}{g} \qquad \text{loss: rounded sheet metal cell} \qquad (5.2)$$

[1] For example, the friction in the sluice gate or the head loss in an open channel flow are taken into account.

[2] This equation also applies to cells made of rounded sheet metal (see Fig. 4.2).

The losses determined by this equation will be too low (since $\gamma \neq 180°$, $c_{2C,u} - u_s \neq w_{3C}$ due to friction or inertia forces). The factor $K = 2$ before the equation must therefore be corrected to $K \approx 1.8$. In the case of full jet deflection, this form of loss only applies to a narrow jet. If the jet is very wide, the jet deflection is no longer possible. The incoming jet can then no longer be fully deflected and is disturbed by the already deflected jet.

It is interesting to note that when $c_{2C,u} = u_s$, the full height H_C becomes effective as a loss. The exit velocity at point A should therefore be higher than the peripheral speed.

One can minimize the loss by reducing the distance between the upper water level and point C, i.e. H_C, and rounding the cell. The aim is to place point C as high as possible. If C is near the upper vertex of the centroid circle, the loss must have a minimum. It can be seen that the variant of the entry into an overshot waterwheel shown in Fig. 3.16 can be significantly improved by reducing δz and shifting δx to more negative values.

In Fig. 5.2, the entry loss for a waterwheel with wooden cells ($R_a = 2$ m, head $H = 4.5$ m, $\delta z = 0.01$ m, speed $n = 8$ min^{-1} and different pressure heights $h_{D2} = 0.2 \ldots 0.5$ m) was determined for different δx. Clear minima of the entry loss heights for different pressure heights h_{D2} or exit velocities are found. Surprisingly, the dependence on the pressure height h_{D2} is comparatively low (0.045 m corresponds to 1% efficiency change), when comparing the minima (3% deviation).

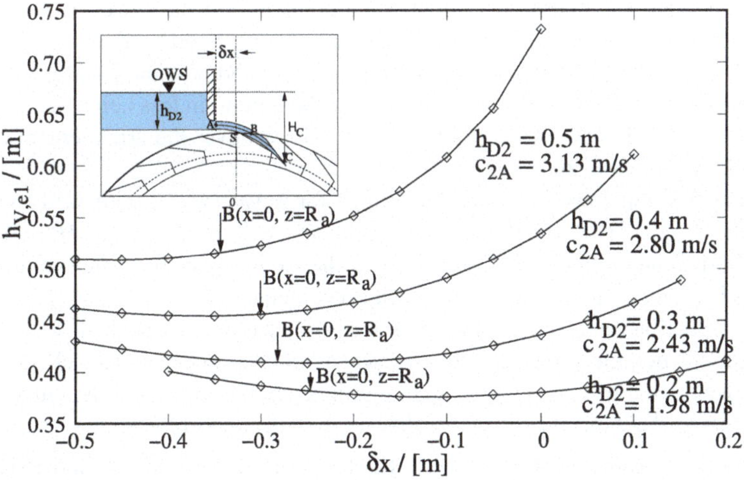

Fig. 5.2 Entry loss at a wooden cell $\gamma = 90°$. The wheel had the following parameters: $R_a = 2$ m, fall height $H = 4.5$ m, $\delta z = 0.01$ m, rotation speed $n = 8$ min^{-1}. The speed has been determined with the equation $c_{2A} = \sqrt{2 \cdot g \cdot h_{D2}}$. Note the start-up behavior!

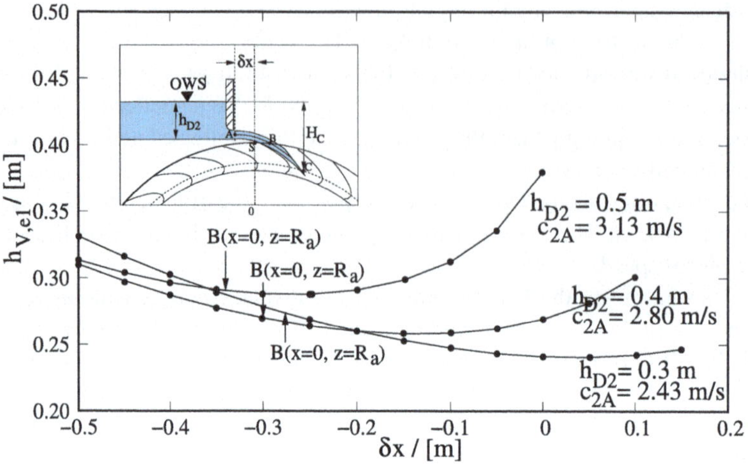

Fig. 5.3 Entry loss at a sheet metal cell, where $\gamma = 160°$ was set. The prefactor was chosen as $K = 1.8$. The wheel had the following parameters: $R_a = 2$ m, fall height $H = 4.5$ m, $\delta z = 0.01$ m, rotation speed $n = 8$ min^{-1}. The speed has been determined with the equation $c_{2A} = \sqrt{2 \cdot g \cdot h_{D2}}$. Note the start-up behavior!

This result also corresponds to the experimental experience of Meerwarth [1] on the overshot waterwheel, who found no significant change in efficiency at different pressure heights $h_{D2} = 0.4$ m, $h_{D2} = 0.2$ m.

Furthermore, the figure shows that when point B is at the apex, the entry loss for all pressure heights is almost minimal. The rule expressed by Bach [2], to place point B at the apex, is thus confirmed. This ensures good efficiency and good start-up behavior. A computational determination of the minimum loss height (δz and δx) should be carried out, as this avoids efficiency losses over the arrangement of the channel for years to come.

In Fig. 5.3, the entry loss was determined for a waterwheel with well rounded sheet metal cells. The wheel data were again: $R_a = 2$ m, fall height $H = 4.5$ m, $\delta z = 0.01$ m and speed $n = 8$ min^{-1}. The change in losses for different pressure heights is even less in the case of rounded sheet metal cells.

The reduction of entry losses through well-rounded or multiple bent sheet metal cells should definitely be applied, as such a rounded sheet metal cell also has advantages at the exit. A 3...6% gain in efficiency is achieved at the entry when using curved sheet metal cells that realize a full deflection of the jet.

*Example 5.1. **(Losses at Water Entry)*** **The losses at the entry for the waterwheel from the last example are to be calculated. Given from the previous example are the circumferential speed at the center of gravity** $u_s = 1.464$ m/s **and the speed** $c_{2C,u} = 4.113$ m/s.

First, the height H_C has to be determined.

$$H_C = R_a + \delta z + m + h_{D2} - z_C$$
$$= 2.0 + 0.1 + 0.005 + 0.4 - 1.486 = 1.019 \text{ m}$$

The entry loss for wooden cells is:

$$h_{V,e1} = H_C - \frac{\left(c_{2C,u} - u_s\right) \cdot u_s}{g}$$
$$= 1.019 - \frac{(4.113 - 1.464) \cdot 1.464}{9.81} = 0.624 \text{ m}$$

Based on the head difference of $H = 4.5$ m, this is already a loss of 13.8%. If one reduces the height H_C by changing $\delta x = -0.4$ the new value is:

$$H_C = R_a + \delta z + m + h_{D2} - z_C$$
$$= 2.0 + 0.1 + 0.005 + 0.4 - 1.770 = 0.735 \text{ m}$$

The pressure height $h_{D2} = 0.4$ m is kept constant and the point C ($C = (x_C = 0.17$ m, $z_C = 1.770$ m)) is moved close to the apex of the center of gravity in the first filled cell.

The speed projected onto the circumferential direction is $c_{2C,u} = 2.935$ m/s. This results in a loss of:

$$h_{V,e1} = 0.735 - \frac{(2.935 - 1.464) \cdot 1.464}{9.81} = 0.515 \text{ m}$$

or related to the head difference of 11.5%. By shifting the entry point, a permanent efficiency gain of 2.3% is achieved. Further potential lies in reducing the distance δz and rounding off the metal sheet of the cells. The distance can be reduced to $\delta z = 0.02$ (consider ice formation).

Now, the optimal ratio between the peripheral speed at the center of gravity and the speed at point C needs to be determined, as well as the dependence of the entry loss on the flow rate.

The second term in Eq. (5.1) must become maximum for a fixed value of H_C in order to keep the entry loss particularly small. The maximum of h is determined with respect to the speed height $\frac{c_{2C,u}^2}{2 \cdot g}$. K is the prefactor for wooden or sheet metal cells, which can vary between 1 and 1.8. The result is:

$$f = \frac{h}{\frac{c_{2C,u}^2}{2 \cdot g}} = \frac{K \cdot \left(c_{2C,u} - u_s\right) \cdot u_s}{g \cdot c_{2C,u}^2/(2 \cdot g)} \quad \text{with} \quad K = 1 \ldots 1.8$$
$$= 2 \cdot K \cdot \left[\frac{u_s}{c_{2C,u}} - \left(\frac{u_s}{c_{2C,u}}\right)^2\right]$$

$$\frac{df}{d\left(\frac{c_{2C,u}}{u_s}\right)} = 1 - 2 \cdot \left(\frac{u_s}{c_{2C,u}}\right) = 0$$

$$\frac{d^2 f}{d\left(\frac{c_{2C,u}}{u_s}\right)^2} = -4 < 0 \quad \text{maximum}$$

The maximum for the function $\frac{h}{\frac{c_{2C,u}^2}{2\cdot g}}$ occurs when:

$$c_{2C,u} = 2 \cdot u_s = 2 \cdot u_a \cdot \frac{R_s}{R_a}$$

This result is independent of the respective cell shape. The optimal ratio between the speed $c_{2C,u}$ and the peripheral speed at the center of gravity is thus determined.

For the example, with a speed of $c_{2C,u} = 4.113$ m, the peripheral speed at the center of gravity should have been $u_s = 2.056$ m/s. The peripheral speed at the outer radius is then comparatively high with $u_a = u_s \cdot \frac{R_a}{R_s} = 2.056 \cdot 2/1.748 = 2.352$ m/s. The entry loss from the previous example is reduced from $h_{V,e} = 0.624$ m to $h_{V,e} = 0.515$ m. A shift of point C towards the apex S also reduces the optimal peripheral speed.

The dependence of the entry loss on the flow rate is determined by the peripheral speed at the center of gravity. The center of gravity radius is dependent on the flow rate, such that losses are reduced with larger flow rates. The change in calculated losses at different flow rates is not very large (refer to Figs. 8.2 and 10.6).

The losses at the outlet increase with the flow rate. Both processes counteract each other, leading to constant efficiencies even at different or very low flow rates.

5.2 Loss Due to the Backlash

The loss due to the backlash on the back of the cell is calculated as follows: First, the amount of water involved in the backlash is determined. The flow rate Q_R is graphically determined from Fig. 5.4, subtracting the backlash-free inflow from the total volume V_z entering the cell.

$$V_R = V_Z - B \cdot A_{Rf} = \frac{Q \cdot 60}{z \cdot n} - B \cdot A_{Rf}$$

For $\frac{z \cdot n}{60}$ cells, there is a backlash on the blade back. The effective flow during the backlash is therefore:

$$Q_R = V_R \cdot \frac{z \cdot n}{60} = Q - \frac{z \cdot n \cdot B \cdot A_{Rf}}{60}$$

In Fig. 5.4, the backlash-free incoming water is represented by the area A_{Rf}. The driving force due to the backlash is:

$$F = \varrho \cdot Q_R \cdot c_{r,u}$$

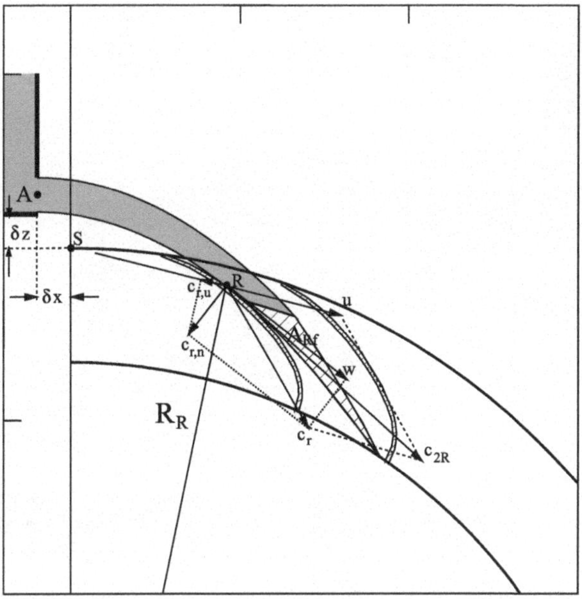

Fig. 5.4 Backlash on the cell

The loss power is then:

$$P_{V,e2} = \omega \cdot R_R \cdot F = \varrho \cdot Q_R \cdot c_{r,u} \cdot u_R$$

where u_R is the circumferential speed at point R. The corresponding loss height is:

$$h_{V,e2} = \frac{Q_R}{Q} \cdot \frac{c_{r,u} \cdot u_R}{g}$$

$$h_{V,e2} = \left(1 - \frac{z \cdot n \cdot B \cdot A_{Rf}}{60 \cdot Q}\right) \cdot \frac{c_{r,u} \cdot u_a}{g} \cdot \frac{R_R}{R_a} \tag{5.3}$$

Depending on how much water enters the cell without backlash, the loss $h_{V,e2}$ occurs. If the entry speed into the waterwheel is too small, the entire flow comes to a stop. The maximum loss due to backlash is then:

$$h_{V,e2} = \frac{c_{r,u} \cdot u_a}{g} \cdot \frac{R_R}{R_a} \tag{5.4}$$

If the cell is curved according to the relative path, the loss due to backlash is zero. The backlash and the filling of the cells cause unpleasant noise when the water is entering the wheel.

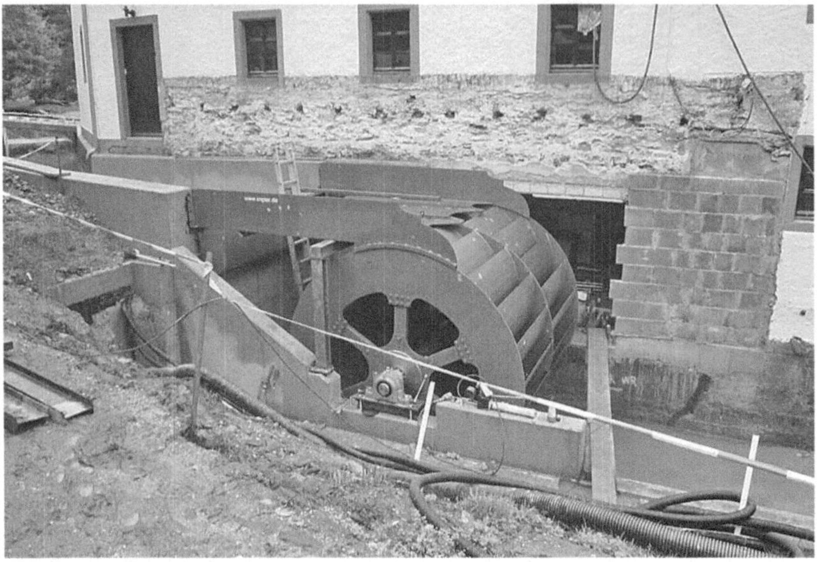

Fig. 5.5 Overshot waterwheel with low head difference and high flow rate. The inlet plate has an exit edge, which is designed to be similar to a sine curve. The water thus enters the wheel at different points, significantly reducing the filling noise, because the air can be easily removed from the cell. In addition, the water's backlash is reduced. Such a wheel in operation is shown in Fig. 6.3. Photo courtesy of Martin Impler GmbH

One way to reduce the noise level is shown in Fig. 5.5. Instead of a straight edge, the inlet plate has a curved shape of the edge. As a result, the water enters the wheel at different points. Ventilation can take place towards the sides and the middle of the jet. This leads to a lower noise level, as individual shocks no longer occur for each cell.

This inlet variant represents a very elegant solution to the problems of entry noise. However, the calculation of the losses becomes more involved.

5.3 Static Exit Losses

The discharge of water begins in the lower third of the waterwheel (see Fig. 5.6). Once the angle φ_1 is reached, the discharge begins. At the angle φ_2, the cell is completely emptied.

The discharged water $Q(h)$ starts from the angle φ_1 continuously over the outer edge of the cells and falls down over a height h to the lower vertex of the waterwheel. This process represents a static energy loss, as the work capacity is lost due to the now missing weight effect.

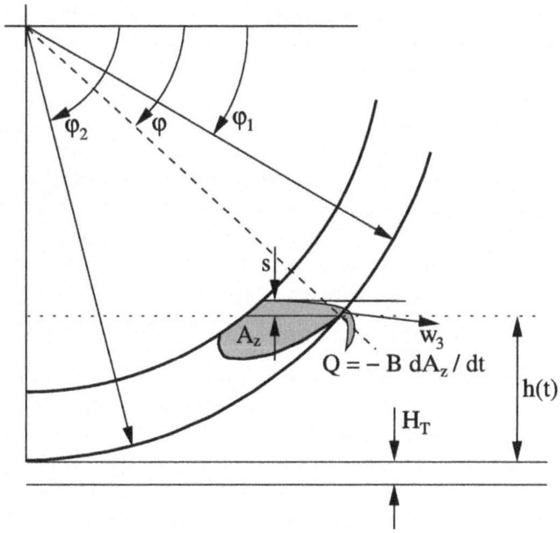

Fig. 5.6 Discharge of water from the cell

A certain overflow height s in the cell is necessary to let the water discharge. If the overflow height is known, it is simple to calculate the lost flow.

Furthermore, the water has a certain kinetic energy when it leaves the cells. This momentum flow is a dynamic loss.

If the waterwheel dips into the tailwater, a loss results from the wheel now wading in the tailwater. The overshot waterwheel should hang freely over the tailwater to avoid wading. This so called free hanging loss depends on the flow rate and the water depth versus discharge curve of the tailwater channel. There is also an overloading loss.

The first three losses mentioned will be calculated in more detail in the following sections. The loss due to overload is only briefly discussed, as it can be avoided by proper design of the wheel. Various numerical or graphical methods for calculating the losses are presented, each leading to the determination of the lost water volume from a single cell.

The determination of the premature discharge from a cell is based on Fig. 5.6.

The water falls down from the cell at time t over a height $h(t)$ down to the lower vertex.[3] The flow that is discharged up to this point is $Q_A(t)$. The instantaneous value of the power lost here is:

$$P(t) = \varrho \cdot g \cdot Q_A(t) \cdot h(t)$$

[3] Here we calculate up to the lower vertex and not to the tailwater level, since the tailwater level is variable over flow rate and weather conditions (weeds in the water, ice etc.).

The average power loss for a whole revolution is obtained from the mean value theorem of integral calculus. We write:

$$P = \varrho \cdot g \cdot \frac{\int_{t_1}^{t_2} Q_A(t) \cdot h(t) \cdot dt}{T}$$

$T = \frac{60}{n}$ is the time duration for a full revolution of the wheel. So far, only the water discharge from a cell has been considered. To get the total loss, the average power loss of one cell is multiplied by the total number of cells z. The total static power loss is now:

$$P_{V,a1} = \frac{z \cdot n}{60} \cdot \varrho \cdot g \cdot \int_{t_1}^{t_2} Q_A(t) \cdot h(t) \cdot dt$$

So far, $Q_A(t)$ and $h(t)$ have not yet been determined. The expression $\frac{z \cdot n}{60}$ corresponds to the cells filled per second. The flow leaving the cell at time t is:

$$\frac{dV_z}{dt} = B \cdot \frac{dA_z}{dt} = -Q_A(t)$$

V_z is the volume of water in the cell. The negative sign in front of $Q_A(t)$ indicates that the volume of water in the cell is decreasing. Since the width of the wheel B is constant, the problem reduces to determining the temporal change in area, since $A_z = V_z/B$ holds.

If one can calculate the change in the available cell area depending on time t, then the prematurely exiting flow $Q_A(t)$ is determined. The cell area covered by water before the water exits is:

$$A_q = \frac{Q \cdot 60}{z \cdot n \cdot B}$$

and serves as a starting value to determine the beginning of the water exit. The height that the water falls down is given by:

$$h(t) = R_a \cdot \left(1 - \sin\left(\omega \cdot t\right)\right)$$

Usually, it is more comfortable to work with angles. If one applies the variable transformation $d\varphi = \omega \cdot dt$, $\varphi = \omega \cdot t$ and inserts the information for $h(t)$ and $Q_A(t)$, one obtains the integral for the average power loss:

$$P_{V,a1} = -\frac{z \cdot n}{60} \cdot \varrho \cdot g \cdot B \cdot \int_{\varphi_1}^{\varphi_2} \frac{dA_z}{d\varphi} \cdot R_a \cdot \left(1 - \sin\varphi\right) \cdot d\varphi$$

At the angle φ_1, the water begins to exit over the cell edge and at φ_2 the cell is emptied.

Since loss heights are more illustrative than powers, the static loss height is used instead of the power loss. One obtains:

$$h_{V,a1} = -\frac{z \cdot n \cdot B}{Q \cdot 60} \cdot \int_{\varphi_1}^{\varphi_2} \frac{dA_z}{d\varphi} \cdot R_a \cdot \left(1 - \sin\,\varphi\right) \cdot d\varphi \tag{5.5}$$

Note that $A_q = \frac{Q \cdot 60}{z \cdot n \cdot B}$ is the area covered with water in a cell before the water exits. The loss height can then also be written as follows:

$$h_{V,a1} = -\frac{\int_{\varphi_1}^{\varphi_2} \frac{dA_z}{d\varphi} \cdot R_a \cdot \left(1 - \sin\,\varphi\right) \cdot d\varphi}{A_q}$$

$h_{V,a1}$ is the first loss which must be taken into account when the water exits. To calculate this loss, knowledge of the function $\frac{dA_z}{d\varphi}$ is necessary, which strongly depends on the respective cell shape and therefore cannot be given analytically. Methods for determining $\frac{dA_z}{d\varphi}$ are derived in the following sections.

The derivative $\frac{dA_z}{d\varphi}$ is negative because the cell area covered by water decreases during water exit. The loss $h_{V,a1}$ is therefore positive.

The static exit loss decreases with increasing rotation speed n, as the covered cell area A_q decreases. The water exit thus begins later, i.e., the angle φ_1 becomes larger. To reduce the static loss, the wheel speed must be increased.

5.4 Dynamic Exit Losses

The instantaneous value of the dynamic power loss is:

$$P = \varrho \cdot Q_A \cdot \left(c_{3,u} - u_a\right) \cdot u_a$$

Thus, for all cells, the power loss relative to one revolution of duration $T = \frac{60}{n}$ is (according to the mean value theorem):

$$P = z \cdot \varrho \cdot \frac{\int_{t_1}^{t_2} Q_A \cdot \left(c_{3,u} - u_a\right) \cdot u_a \cdot dt}{T}$$

Analogous to the static loss, instead of times, it is easier to work with angles:

$$P = -\frac{z \cdot n}{60} \cdot \varrho \cdot B \cdot \int_{\varphi_1}^{\varphi_2} \frac{dA}{d\varphi} \cdot \left(c_{3,u} - u_a\right) \cdot u_a \cdot d\varphi$$

where the relationship $Q_A = -B \cdot \omega \cdot \frac{dA_z}{d\varphi}$ already used in the last section was applied. The loss height[4] is now:

$$h_{V,a2} = -\frac{z \cdot n \cdot B}{60 \cdot Q} \cdot \int_{\varphi_1}^{\varphi_2} \frac{dA}{d\varphi} \cdot \frac{\left(c_{3,u} - u_a\right) \cdot u_a}{g} \cdot d\varphi$$

$$h_{V,a2} = -\frac{\int_{\varphi_1}^{\varphi_2} \frac{dA}{d\varphi} \cdot \left(c_{3,u} - u_a\right) \cdot u_a \cdot d\varphi}{g \cdot A_z} \tag{5.6}$$

Only $c_{3,u}$, the circumferential component of vector $\vec{c}_3$ is still unknown, assuming that $\frac{dA_z}{d\varphi}$ is known (for the calculation of this derivative, see the following sections). The absolute discharge velocity c_3 can be calculated using the cosine theorem (see Fig. 5.7):

$$c_3 = \sqrt{u_a^2 + w_3^2 - 2 \cdot u_a \cdot w_3 \cdot \cos\left(90^\circ - \varphi\right)}$$

and the circumferential component of the discharge velocity is:

$$c_{3,u} = c_3 \cdot \cos\alpha_3$$

The angle α_3, which the exit velocity c_3 encloses with the circumferential velocity, results from repeated application of the cosine theorem:

$$\cos\alpha_3 = \left(\frac{u_a^2 + c_3^2 - w_3^2}{2 \cdot u_a \cdot c_3}\right)$$

The equations assume that the exit from the cell is horizontal. The effect of the centrifugal force is thus not taken into account.[5]

---

[4] The loss height is negative, so one might think that there is an energy gain to record. However, the change in area is negative $\frac{dA_z}{d\varphi} < 0$, so that actually, an energy loss is calculated.

[5] See Figs. 1.2 and 4.1 and the text there.

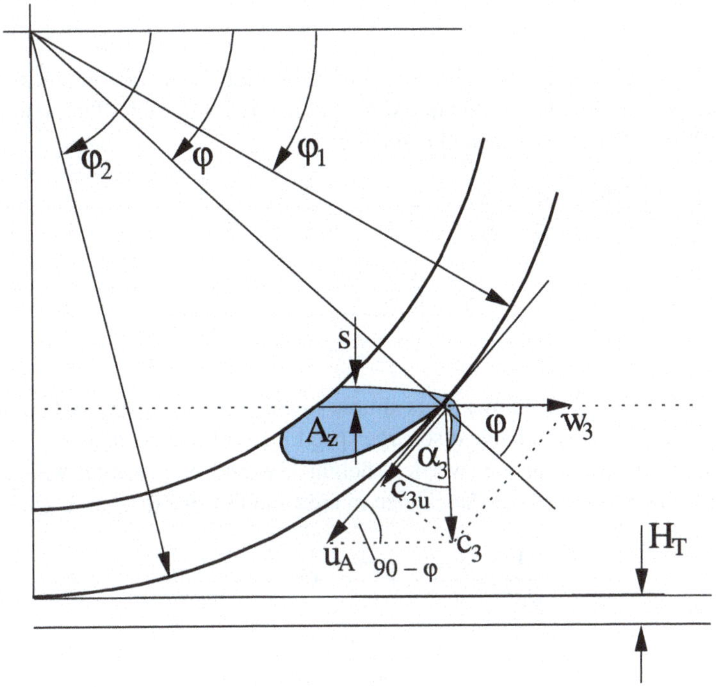

Fig. 5.7 Water exiting the cell of an overshoot waterwheel. The components of the exit velocity c_3 are shown as a velocity triangle

In the calculations, only the expression for the relative exit velocity w_3 is now missing. This will now be determined. From the continuity equation for water discharge, we obtain:

$$B \cdot \frac{dA_z}{dt} = B \cdot \omega \cdot \frac{dA_z}{d\varphi} = -Q_A = -w_3 \cdot A_{out} = -w_3 \cdot \frac{2}{3} \cdot s \cdot B$$

$A_{out} = \frac{2}{3} \cdot s \cdot B$ is the discharge area through which the water leaves the cell and s is the overfall height necessary for the discharge (see Fig. 5.7).[6] The relative speed is:

[6] The correctness of the approach for the area and the speed can be checked as follows (see Bollrich [3], derivation for the overflow weir and reference [4]). The energy height is $h_E = s = z + \frac{w_3^2}{2g}$ (the centrifugal force is neglected), so $w_3(z) = \sqrt{2 \cdot g \cdot (s - z)}$. The discharge over the cell exit edge is now calculated by:

$$dQ = w_3(z) \cdot dA = B \cdot \sqrt{2 \cdot g \cdot (s - z)} dz$$

$$w_3 = \sqrt{2 \cdot g \cdot s}$$

This relative discharge speed is somewhat too high, as no coefficient for speed reduction was assumed. If both equations are inserted into each other, one obtains the overflow height s over the cell exit edge:

$$-B \cdot \frac{2 \cdot \pi \cdot n}{60} \cdot \frac{dA_z}{d\varphi} = \frac{2}{3} \cdot \sqrt{2 \cdot g \cdot s} \cdot s \cdot B \quad \text{or}$$

$$s = \left(\frac{1}{\frac{2}{3} \cdot \sqrt{2 \cdot g}} \cdot \left[-\frac{2 \cdot \pi \cdot n}{60} \cdot \frac{dA_z}{d\varphi} \right] \right)^{2/3}$$

If the function $\frac{dA_z}{d\varphi}$ is known, it is simple to calculate the overflow height s and thus also the respective relative discharge speed $w_3(\varphi)$ for each angle φ.

Similarly, determining the overflow height s would be possible, which would then allow conclusions about the change in area and thus the discharge Q_A.

$$\frac{dA_z}{d\varphi} = -\frac{2}{3} \cdot \sqrt{2 \cdot g} \cdot s^{3/2} \cdot \frac{60}{2 \cdot \pi \cdot n}$$

$$Q_A = B \cdot \frac{2 \cdot \pi \cdot n}{60} \cdot \frac{dA_z}{d\varphi}$$

This results in two methods for determining the dynamic and static losses (determining $\frac{dA_z}{d\varphi}$ or determining s). The exit speed c_3 is determined by the relative speed $w_3(\varphi)$ and the circumferential speed u_a.

With the knowledge of the magnitudes of the vectors c_3, u_a, w_3, we can now determine α_3 and thus also the dynamic power loss or loss height $h_{V,a2}$.

It is desirable to make the exit edge from the cell particularly sharp-edged. The outflow is then inhibited. The overflow height s will be higher than with a rounded exit edge and the speed w_3 will decrease. If w_3 decreases, the dynamic loss decreases, as $c_{3,u}$ becomes smaller.

The calculation of the dynamic loss is quite complex. However, the calculation is inaccurate because the centrifugal force acting on each individual cell is neglected. The dynamic head loss is overestimated with the equations above.

$$Q = B \cdot \int_0^s \sqrt{2 \cdot g \cdot (s - z)} dz = \left[-\frac{2}{3} \cdot \sqrt{2 \cdot g} \cdot B \cdot (s - z)^{2/3} \right]_0^s$$

$$Q = \frac{2}{3} \cdot \sqrt{2 \cdot g \cdot s} \cdot B \cdot s = w_3 \cdot A \quad \text{where}$$

$$A_{out} = \frac{2}{3} \cdot s \cdot B \quad \text{and} \quad w_3 = \sqrt{2 \cdot g \cdot s}$$

The reduction of speed due to friction is neglected in these equations.

Meerwarth [1] has found that the dynamic loss can be approximately described by:

$$h_{V,a2} = 0.8 \cdot \frac{u_a^2}{2 \cdot g} \tag{5.7}$$

5.5 Determination of Exit Losses: Method 1

The method is based on the determination of the area change $\frac{dA}{d\varphi}$. The wooden cell from the last example is used for the investigation. In Fig. 5.8, this wooden cell is shown rotated by 90°.

The procedure to determine the losses is:

1. The cell of the waterwheel should be drawn as shown in Fig. 5.8. The coordinates of point E should be $E(x_E = R_a, z_E = 0)$. If the coordinates of the cell are available, then a rotation of the coordinates $P(x_P, z_P)$ into the point $P' = (x_{P'}, z_{P'})$ via the equations:

$$x_{P'} = x_P \cdot \cos\gamma - z_P \cdot \sin\gamma$$

$$z_{P'} = x_P \cdot \sin\gamma + z_P \cdot \cos\gamma$$

   is easily possible. Here, γ is the rotation angle.
2. As a second step, determine the angle φ_2, at which the exit is completed. For the shown wooden cell, one obtains e.g.:

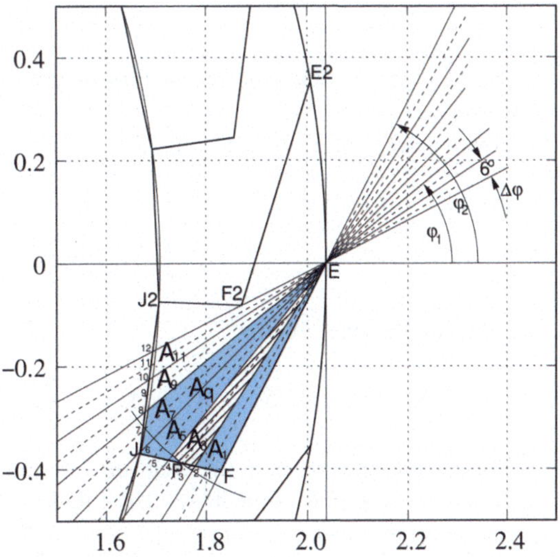

Fig. 5.8 Wooden cell for calculation of volume change and exit losses

$$\varphi_2 = \arctan \frac{z_E - z_F}{x_E - x_F} = 62^\circ$$

From this angle, the cell is emptied. For sheet metal cells, determine the tangent at point E and then the angle φ_2. Additionally, determine the area occupied by the water in the cell:

$$A_z = \frac{Q \cdot 60}{n \cdot z \cdot B}$$

3. Now draw a bundle of straight lines through point E. The lines must be offset by an angle $\Delta\varphi/2$:

$$\varphi = \varphi_2 - \frac{\Delta\varphi}{2} \cdot i \qquad i = 0, 1, 2 \ldots n \qquad \text{where} \qquad \Delta\varphi = 2^\circ \ldots 6^\circ$$

against each other. For a wooden cell, it is as shown in Fig. 5.8.
4. Determine the intersection points of the lines with the cell (points 1 to 12). Then calculate the lengths between the respective point and point E (e.g. determine the length $\widehat{EP_3}$). If the coordinates of the respective point, e.g. P_3, are known, then the distance between E and P_3 is:

$$\widehat{EP_3} = \sqrt{(x_E - x_{P_3})^2 + (z_E - z_{P_3})^2}$$

5. Calculate the areas A_1, A_3, $A_5 \ldots$ (for example, see the hatched area A_3 in Fig. 5.8):

$$A_1 = \frac{\pi \cdot (\widehat{EP_1})^2}{360^\circ} \cdot \Delta\varphi, \quad A_3 = \frac{\pi \cdot (\widehat{EP_3})^2}{360^\circ} \cdot \Delta\varphi, \quad A_5 = \frac{\pi \cdot (\widehat{EP_5})^2}{360^\circ} \cdot \Delta\varphi \ldots$$

These areas are the circular sectors of the circles with the radii $\widehat{EP_1}$, $\widehat{EP_3}$, $\widehat{EP_5} \ldots$.
6. Now the area sums are determined:

$$S_1 = \sum_{n=1}^{i} A_i$$

For example, one obtains

$$S_1 = A_1 \qquad S_3 = A_1 + A_3 \qquad S_5 = A_1 + A_3 + A_5 \qquad \ldots$$

The comparison between the area sums S_i and the covered cell area $A_z = \frac{Q \cdot 60}{n \cdot z \cdot B}$ shows up to which point the cell is filled for the flow Q and a fixed speed n just before the water outlet (hatched area in Fig. 5.8). This gives the angle φ_1, at which the water outlet begins.

7. Determine the speeds associated with the areas:

$$n_1 = \frac{Q \cdot 60}{S_1 \cdot z \cdot B} \qquad n_3 = \frac{Q \cdot 60}{S_3 \cdot z \cdot B} \qquad n_5 = \frac{Q \cdot 60}{S_5 \cdot z \cdot B} \qquad \ldots$$

These speeds correspond to the area $S_1 = A_1$, $S_3 = A_1 + A_3$, $S_5 = A_1 + A_3 + A_5$ $S_7 = \ldots$, which the water occupies before the cell content $A_z = \frac{Q \cdot 60}{n \cdot z \cdot B}$ is emptied.

8. After that, the respective derivatives must be determined (point P_3):

$$\frac{dA_1}{d\varphi} \approx \frac{\Delta A_1}{\Delta\varphi} = \frac{0 - A_1}{\Delta\varphi} = -\frac{A_1}{\Delta\varphi} = -\frac{\pi \cdot (\widehat{EP_1})^2}{360°}$$

$$\frac{dA_3}{d\varphi} \approx \frac{\Delta A_3}{\Delta\varphi} = \frac{A_1 - (A_1 + A_3)}{\Delta\varphi} = -\frac{A_3}{\Delta\varphi} = -\frac{\pi \cdot (\widehat{EP_3})^2}{360°}$$

$$\frac{dA_5}{d\varphi} \approx \frac{\Delta A_5}{\Delta\varphi} = \frac{(A_1 + A_3) - (A_1 + A_3 + A_5)}{\Delta\varphi} = -\frac{A_5}{\Delta\varphi} = -\frac{\pi \cdot (\widehat{EP_5})^2}{360°}$$

$$\vdots$$

The respective derivative is negative because the water-covered area decreases with increasing angle. The derivative is proportional to the length change $\widehat{EP_n}$.

9. At the end the individual static losses are determined:

$$h_{V,a1,i} = \frac{1}{S_i} \cdot \sum_{i=1}^{n} A_i \cdot R_a \cdot \left[1 - \sin\left(\varphi_2 - i \cdot \frac{\Delta\varphi}{2}\right)\right] \qquad i = 1, 3, 5 \ldots n$$

Here, the integral[7] (5.5) was rewritten into a sum. For example, the following results are obtained:

$$h_{V,a11} = \frac{1}{S_1} \cdot \left[A_1 \cdot R_a \cdot \left(1 - \sin\left(\varphi_2 - 1 \cdot \frac{\Delta\varphi}{2}\right)\right)\right] \quad \text{loss for} \quad A_1 = A_z$$

$$h_{V,a13} = \frac{1}{S_3} \cdot \left[A_1 \cdot R_a \cdot \left(1 - \sin\left(\varphi_2 - 1 \cdot \frac{\Delta\varphi}{2}\right)\right)\right.$$
$$\left. + A_3 \cdot R_a \cdot \left(1 - \sin\left(\varphi_2 - 3 \cdot \frac{\Delta\varphi}{2}\right)\right)\right] \quad \text{loss for} \quad A_1 + A_2 = A_z$$

$$\vdots$$

The previously determined area derivatives and the sums S_i were used. This then determines the total static loss for the respective associated speed.

[7] The integrand of the integral (5.5) is assumed to be constant at point i (interpolation quadratures–rectangle rule).

10. In the same way, the individual dynamic losses are obtained:

$$h_{V,a1,i} = \frac{1}{S_i} \cdot \sum_{n=1}^{i} \cdot A_i \cdot \frac{(c_{3,i}^2 - w_3^2)}{2 \cdot g}$$

In the equation above, $c_{3u,i}$ is to be determined as follows:

$$c_{3,i} = \sqrt{u_a^2 + w_{3,i}^2 - 2 \cdot u_a \cdot w_{3,i} \cdot \cos(90^\circ - \varphi_i)}, \text{ where}$$

$$w_{3,i} = \sqrt{2 \cdot g \cdot s_{3,i}} \quad \text{and}$$

$$s_i = \left(\frac{1}{\frac{2}{3} \cdot \sqrt{2 \cdot g}} \cdot \left[-\frac{2 \cdot \pi \cdot n}{60} \cdot \frac{dA_z}{d\varphi} \right] \right)^{2/3}$$

For the first two losses, for example, one obtains:

$$h_{V,a21} = \frac{1}{S_1} \cdot \left[A_1 \cdot \frac{(c_{3u,1} - u_a)}{g} \cdot u_a \right]$$

$$h_{V,a23} = \frac{1}{S_3} \cdot \left[A_1 \cdot \frac{(c_{3u,1} - u_a)}{g} \cdot u_a + A_3 \cdot \frac{(c_{3u,3} - u_a)}{g} \cdot u_a \right]$$

$$\vdots$$

As Meerwarth [1] has already determined, the dynamic loss can be well represented by the equation:

$$h_{V,a2} = 0.8 \cdot \frac{u_a^2}{2 \cdot g} \quad \text{or through}$$

$$h_{V,a2,i} = 0.8 \cdot \frac{\pi^2 \cdot R_a^2 \cdot n_i^2}{1800 \cdot g}$$

without having to accept the high computational effort of the above formulas. The dynamic total loss can also be assigned to the respective speed n_i.

11. The losses associated with the speeds (sum areas) are:

$$S_1 \rightarrow n_1 \quad \rightarrow$$

$$h_{V,a1} = h_{V,a11} \quad \text{and} \quad h_{V,a2} = h_{V,a21}$$

$$S_3 \rightarrow n_3 \quad \rightarrow$$

$$h_{V,a1} = h_{V,a13} \quad \text{and} \quad h_{V,a2} = h_{V,a23}$$

$$S_5 \rightarrow n_5 \quad \rightarrow$$

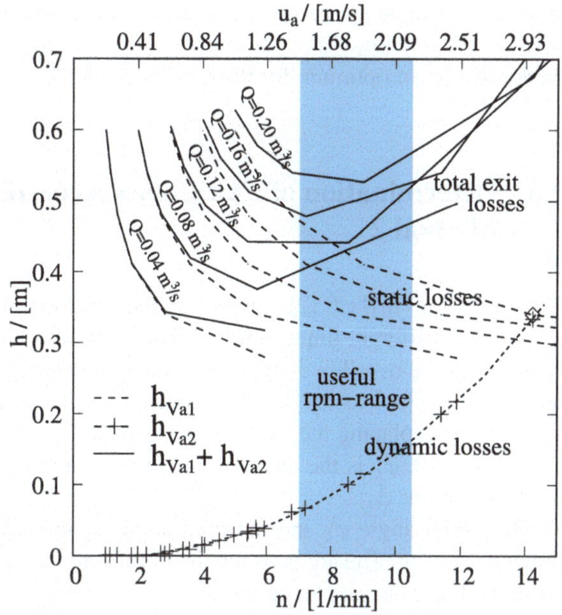

Fig. 5.9 Static and dynamic losses of the wood cell in Fig. 5.8

$$h_{V,a1} = h_{V,a15} \quad \text{and} \quad h_{V,a2} = h_{V,a25}$$

$$\vdots$$

Intermediate values of losses can be determined by linear interpolation of neighboring values. Interpolation methods following Newton or Lagrange can be found e.g. in Bronstein [5].

12. If the losses are tabulated over the speed, then the desired relationship between the speed and the dynamic or static losses for a certain flow rate Q can be shown.
13. The procedure is now repeated for different flow rates Q. In this case, the cell area A_z changes with the flow rate at a constant speed. In this simple way, one obtains the losses for different flow rates and speeds. This is shown in Fig. 5.9 for the wood cell shown in Fig. 5.8. The curves in the representation become "smoother" if one uses smaller changes of the angle $\Delta\varphi$.

Here, the advantage of using angles becomes clear. We achieve independence from time and angular velocity, since $\varphi = \omega \cdot t$ applies. We can therefore determine the losses for very different operating conditions (flows and speeds) as shown in Fig. 5.8.

As can be seen from Fig. 5.9, the static losses decrease with increasing speed n and decreasing flow rate Q. This is due to the greater filling of the cell at lower speed or increased flow rate. The water exit therefore starts earlier.

The dynamic losses, on the other hand, increase with the speed, since c_3 increases with it. Adding up both loss heights $h_{V,a1}$ and $h_{V,a2}$, one finds that there is a minimum at higher speeds. A peripheral speed that is too low therefore leads to smaller efficiencies. Since the entry losses and the exit losses have a minimum, there must be an optimum for the rotational speed.

5.6 Determination of Exit Losses According to Bach: Method 2

The procedure of Bach [2] is based on the graphical determination of the "overfall height" s at different times (angles). The method is shown in Fig. 5.10 for wooden cells. From the overflow height, one can then determine the area change $\frac{dA}{d\varphi}$ or the outflow Q_A. The loss heights $h_{V,a1}$ and $h_{V,a2}$ can be calculated with the Eqs. (5.5) and (5.6) by replacing the integrals with sums.

In Fig. 5.10, φ_1 is the angle at which the water begins to exit the cells. This is marked as point A.

The initial angle φ_1 and the final angle φ_2 are determined as explained in the previous section. The arc between the initial and final angle is divided into equal arc sections. For example, let an arc piece $AC = \frac{2 \cdot \pi \cdot R_a}{360°} \cdot \gamma_t$ be given, where γ_t is the division angle.

To determine the amount of water still contained in the cell space when the cell has reached point C, draw another cell halfway between A and C. Then, draw a

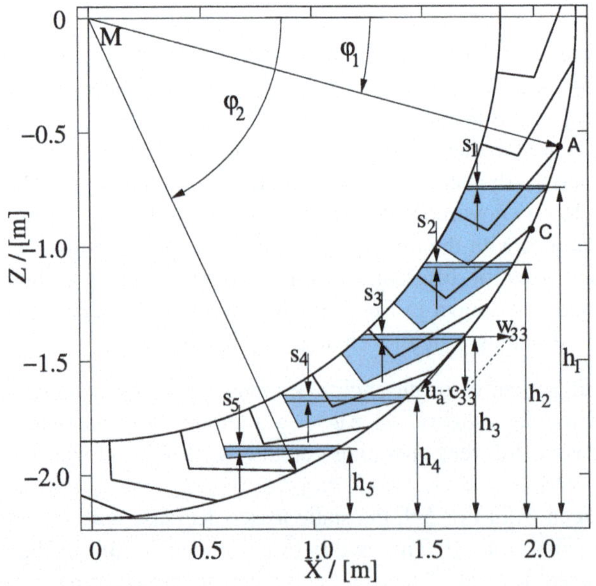

Fig. 5.10 Graphical determination of exit losses

horizontal line which intersects the cell exit edge. In addition, the area content of the previous cell (at point A) is determined and entered into the cell between A and C. One can now read off the corresponding overflow height s_1, which must form in the cell to allow the flow:

$$Q_{A,1} = \frac{2}{3} \cdot \sqrt{2 \cdot g \cdot s_1} \cdot s_1 \cdot B = -B \cdot \omega \cdot \left(\frac{\Delta A}{\Delta \varphi}\right)_1$$

to be discharged. $Q_{A,1}$ falls by the height h_1. With the overfall height s, one can now also calculate the relative exit velocity:

$$w_{3.1} = \sqrt{2 \cdot g \cdot s_1}$$

Again, we now know the quantities s_1 or w_{31}, h_1 and $\left(\frac{\Delta A}{\Delta \varphi}\right)_1$ or $Q_{A,1}$, so that the static and dynamic loss for this cell can be calculated. For all further cells, the procedure is carried out in the same way. The sum formulas for the static and dynamic loss can be calculated in the same way as explained in the previous procedure. The total loss is obtained by summing the two loss heights.

The procedure is somewhat cumbersome and offers no significant advantages over the first method. The only advantage is that one can get a picture of the effect of the centrifugal force. For this, instead of the horizontal mirrors, the circular level surfaces in the cells are drawn. Bach [2] presents this procedure in detail in his work.

5.7 Free Hanging and Submersion

There is also a loss of power due to the free hanging H_T of the waterwheel above the lower water level:

$$P_{v,a3} = \varrho \cdot g \cdot H_T \cdot Q$$

The associated loss height is:

$$h_{V,a3} = H_T \tag{5.8}$$

The head loss because of free hanging is not constant, as the water depth h in the tail water varies depending on the flow rate available in the tailwater channel. Due to the usually very small river bed slopes in the tailwater, the water depth changes significantly with flow rate here. When the waterwheel dips into the underwater, wading losses occur. Meerwarth [1] has indicated three wading losses:

1. losses due to the lifting of water in the cells due to the resulting negative pressure in the respective closed cell space. Assuming that the volume of air is not compressed or expanded, the heights h_2 to h_4 can be determined via the area

calculation formulas ($V_1 = B \cdot A_1$). This loss can be avoided if the cells have an opening towards the wheel bottom (This causes losses at the water entry.).
2. losses due to the flow resistance of the waterwheel and
3. losses due to surface resistance (friction) when immersing into the underwater.

An idea of the wading losses can be obtained as follows: Imagine a straight blade that dips into the (assumed to be still) underwater with the average peripheral speed. The force acting on the blade at point P is:

$$F = \varrho \cdot Q_1 \cdot u \quad \text{with}$$

$$u = u_a \cdot \frac{R_P}{R_a} = u_a \cdot \frac{R_a - H_T/2}{R_a}$$

The submerged waterwheel is shown in Fig. 5.11. The water flow Q_1 is determined from the figure as:

$$Q_1 = B \cdot H_T \cdot u_a \cdot \frac{R_a - H_T/2}{R_a}$$

From the counterforce F, the counter moment M, and from the equation above the loss power is determined:

$$M = F \cdot R_P = \varrho \cdot Q_1 \cdot u_a \cdot \frac{R_a - H_T/2}{R_a} \cdot R_P \quad \text{and thus}$$

$$P_{V,a4} = M \cdot \omega = \varrho \cdot B \cdot H_T \cdot u_a^3 \cdot \left(\frac{R_a - H_T/2}{R_a}\right)^3$$

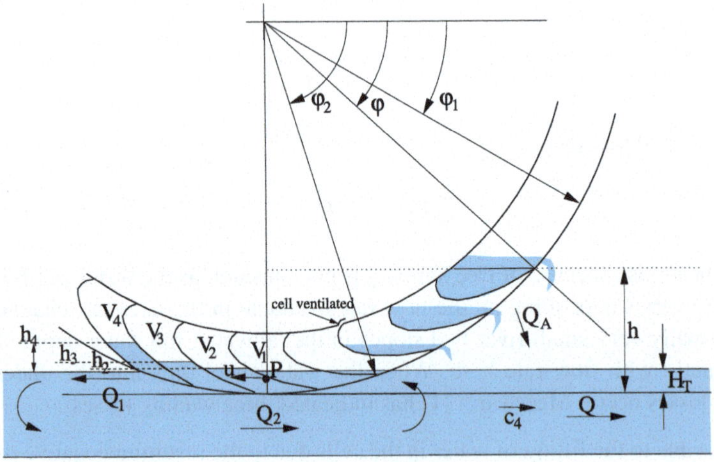

Fig. 5.11 Wading of the waterwheel

If this loss power is written as loss height, the result is:

$$h_{V,a4} = \frac{B \cdot H_T}{g \cdot Q} \cdot u_a^3 \cdot \left(\frac{R_a - H_T/2}{R_a}\right)^3 \tag{5.9}$$

The wading loss increases cubically with the peripheral speed and is proportional to the area $B \cdot H_T$. Furthermore, it is important that when the waterwheel wades, the height $h(t)$ of the static exit loss $h_{V,a1}$ has to be corrected. This correction is particularly simple if the water completely empties the cell before reaching the underwater level. Then a gain of head difference of H_T has to be taken into account.

The loss calculated here is only a rough approximation. No consideration was given to water vortexes under the wheel and the compression and expansion of air in the cells. To reduce the loss due to free hanging, one must determine the water depth versus flow rate curve of the channel (see Manning's formula (2.1)). However, a more accurate method is the simultaneous measurement of the underwater level and the flow rate.

For the largest flow to be discharged by the waterwheel, the underwater level should touch the lower vertex of the waterwheel. For flows greater than the nominal flow, the wheel wades. Meerwarth [1] recommends a wading of the wheel up to $H_T = 0.05 \cdot R_a$ to keep the free hanging loss small. However, this was only checked using a small model wheel.

These results can also be transferred to a large wheels, as the measurement results for wading experiments Carl R. Weidner 1913, University of Wisconsin [6] show.

A free hanging of $H_T = -0.05 \ldots - 0.1$ m, as often found in older waterwheel literature, is not recommended, as this loss of head should not be permanently accepted. If a backwater is observed during flooding conditions from the main arm of the stream or river, one should determine the underwater level with an accurate level measurement in advance.

One then needs to determine on how many days the backwater occurs and whether the waterwheel really needs to be positioned higher because of this. The channel in the underwater should be as wide and deep as possible, to keep the change of the water level at different flow rates small.

5.8 Losses Due to Overflow of the Wheel

There is an additional loss term to be considered if the waterwheel is overloaded by a overly high flow rate $Q = Q_{overl} + Q_L$. In this case, the flow rate exceeds both the design flow rate as well as the limiting overload flow rate Q_{overl}. The limiting flow rate has already been derived in Eq. (4.19).

This causes a reduction in the efficiency of the overshot waterwheel due to an additional leakage flow which can be determined as follows:

Fig. 5.12 Overloading of a overshot waterwheel. The wheel is running too slow. The wheel was later replaced by a new wheel designed by the author. Refer to Figs. 3.20, 3.21, 3.22 (With kind permission of Tobias Gappisch Wasserkraft- und Anlagentechnik, Göda)

$$\eta_{WW,total} = \eta_{WW} + \left(\frac{Q_{overl}}{Q} - 1\right), \quad \text{if } Q > Q_{overl}$$

$$\eta_{WW,total} = \eta_{WW}, \quad \text{if } Q \leq Q_{overl} \tag{5.10}$$

The leakage flow is shown in Fig. 4.1 (upper right corner).

The efficiency of the overflowed waterwheel decreases very quickly due to the leakage current $Q_L = Q - Q_{overl}$ if the flow rate Q is increased beyond Q_{overl}. Figure 5.12 shows a side view of an overflown waterwheel under such conditions.

References

1. Meerwarth KW (1935) Experimentelle und theoretische Untersuchungen am oberschlächtigen Wasserrad. PhD thesis, Technische Hochschule Stuttgart
2. Bach Cv (1886) Die Wasserräder, 1st edn. Verlag von Konrad Wittwer, Stuttgart
3. Bollrich G, Preißler G (1992) Technische Hydromechanik Band 1, 3rd edn. Verlag für Bauwesen, Berlin
4. Nuernbergk DM (2005) Wasserräder mit Kropfgerinne – Berechnungsgrundlagen und neue Erkenntnisse, 1st edn. Verlag Moritz Schäfer, Detmold
5. Bronstein IN, Semendjajew KA, Musiol G, Mühlig H (2000) Taschenbuch der Mathematik, 5th edn. Verlag Harri Deutsch, Frankfurt am Main
6. Weidner CR (1913). Theory and test of an overshot water wheel. Bull Wis Univ Eng Ser Madison, Wisconsin 7(529):157–254

Chapter 6
Losses and Efficiency

6.1 Work Capacity of Water

What energy can be extracted from flowing water to drive a hydraulic machine, such as a waterwheel? A flowing liquid possesses energy in three forms [1]:

1. Potential energy $F_G \cdot z = m \cdot g \cdot z$,
2. Pressure energy $V \cdot p = m \cdot p/\varrho$ and
3. Kinetic energy $m \cdot c^2/2$.

The total hydraulic energy in a frictionless liquid is:

$$W = m \cdot g \cdot \left(z + \frac{p}{\varrho \cdot g} + \frac{c^2}{2 \cdot g}\right)$$

The hydraulic energy is the same at every point, as frictionlessness implies lossless flow.

If the energy equation is expressed as energy height h_E, the Bernoulli equation, named after Daniel Bernoulli,[1] results:

$$\frac{W}{m \cdot g} = h_E = \left(z + \frac{p}{\varrho \cdot g} + \frac{c^2}{2 \cdot g}\right) = const$$

In a stationary flow of a frictionless liquid, the energy height h_E for any reference point is composed of the position z, the pressure height $h_D = p/(\varrho \cdot g)$ and the velocity height $h_{kin} = c^2/(2 \cdot g)$.

[1] Daniel Bernoulli, ∗29.01.1700 †27.3.1782, Swiss mathematician and physicist.

D. M. Nuernbergk, *Overshot Waterwheels*, Springer Tracts in Mechanical Engineering, https://doi.org/10.1007/978-3-031-95385-9_6

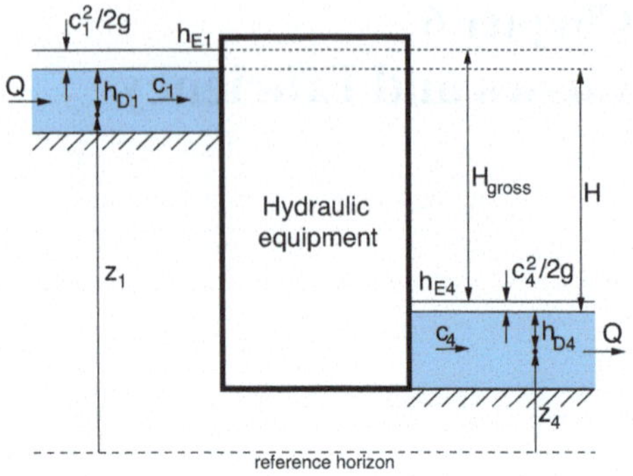

Fig. 6.1 Energy relationships in a hydropower system

In a lossless, frictionless liquid, the energy height remains constant. In the direction of flow, only the proportions of position, pressure height, and velocity height change.

For two different flow cross-sections 1 and 4, the following applies:

$$h_E = z_1 + \frac{p_1}{\varrho \cdot g} + \frac{c_1^2}{2 \cdot g} = z_4 + \frac{p_4}{\varrho \cdot g} + \frac{c_4^2}{2 \cdot g}$$

i.e., the energy line remains constant without friction losses. Without considering friction losses, the Bernoulli equation sometimes yields unphysical results (longer flow paths without friction). For processes where the internal losses can be neglected, it already provides technically useful results in the form given above (short flow paths) (Fig. 6.1).

This form of the Bernoulli equation only applies to a stationary flow. The energy that can be extracted from the water between two points by a hydraulic machine is:

$$W_{hydr} = m \cdot g \cdot \left((z_1 - z_4) + \frac{p_1 - p_4}{\varrho \cdot g} + \frac{c_1^2 - c_4^2}{2 \cdot g} \right)$$

Energy is extracted from the water and the energy line jumps by $H_{gross} = h_{E1} - h_{E2}$ at the hydraulic machine. This energy can be extracted from the water by changing the pressure, the geodetic height, or the speed. The fall height H is the difference in water levels $H = z_1 + \frac{p_1}{\varrho \cdot g} - z_4 - \frac{p_4}{\varrho \cdot g}$, where $p_1 = p_4 = p_{amb}$ is the air pressure.

Therefore, the energy that can be extracted from the water by a waterwheel is equal to:

$$W_{hydr} = m \cdot g \cdot \left(H + \frac{c_1^2 - c_4^2}{2 \cdot g} \right)$$

The derivative of the energy with respect to time gives the theoretically available power for a steady flow state:

$$P_{th} = \frac{dW_{hydr}}{dt}$$
$$= \dot{m} \cdot g \cdot \left(H + \frac{c_1^2 - c_4^2}{2 \cdot g}\right) = \varrho \cdot g \cdot \dot{V} \cdot \left(H + \frac{c_1^2 - c_4^2}{2 \cdot g}\right)$$
$$= \varrho \cdot g \cdot Q \cdot \left(H + \frac{c_1^2 - c_4^2}{2 \cdot g}\right) = \varrho \cdot g \cdot Q \cdot H_{gross}$$

$$P_{th} \approx \varrho \cdot g \cdot Q \cdot H = 9800 \cdot Q \cdot H \quad \text{in [W]} \tag{6.1}$$

$Q = \dot{V} = \frac{dV}{dt}$ is the water flow (volume flow, throughput), g is the gravitational acceleration[2] and ϱ is the density of the water. It is assumed that the flow process is steady, i.e., the velocities c_1^2, c_4^2 and the fall height H are constant.

The difference in kinetic energies in the inflow and outflow areas are usually not usable, as they are approximately equal. It is therefore often assumed that the geodetic fall height is equal to the gross head height $H = H_{gross}$. The theoretical power P_{th} is directly proportional to the head difference and the flow rate. If the efficiency η_{WW} of the waterwheel is known, the mechanical power that can be extracted from the waterwheel at the waterwheel shaft is:

$$P_N = \eta_{WW} \cdot P_{th} \tag{6.2}$$

An essential part of the design of waterwheels is therefore the calculation of the efficiency. From the course of the head difference, the flow duration line and the efficiency, the annual energy yield E can be calculated. This value is the input parameter for profitability calculations.

6.2 Losses

The losses (Fig. 6.2), which were calculated as loss heights in the previous chapter, are:

1. Losses due to contraction and friction in the sluice gate, in the tongue inlet or losses due to different water depths in the inlet channel, when supplied via an open channel.

[2] The gravitational or earth acceleration changes with geographical latitude due to the centrifugal force, from the equatorial value $g = 9.78$ m/s^2 up to $g = 9.832$ m/s^2 at the poles. Therefore, the standard gravity $g = 9.80665$ m/s^2 should be used. In this work, the rounded value $g = 9.81$ m^2/s is used for simplicity.

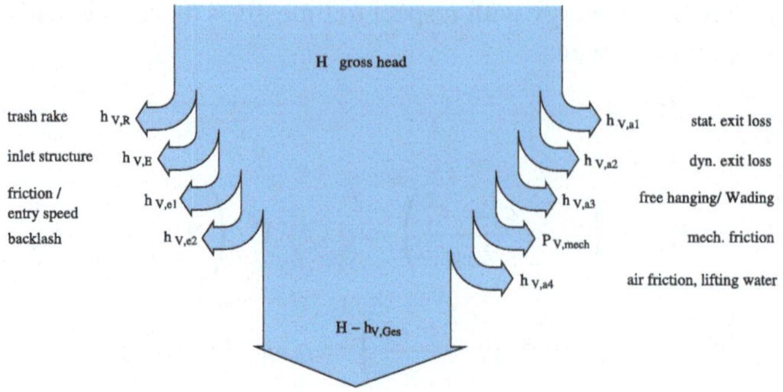

Fig. 6.2 Losses on the overshot waterwheel represented as loss heights

2. Losses due to the imperfect transfer of impulse to the cell. The sum of both losses is accomplished by Eq. (5.1). The combined losses are:

$$h_{V,e1} = H_C - K \cdot \frac{\left(c_{2C,u} - u_s\right) \cdot u_s}{g} \quad \text{with} \quad K = 1 \ldots 1.8$$

depending on the rounding of the cells.

3. The loss due to recoil:

$$h_{V,e2} = \left(1 - \frac{z \cdot n \cdot B \cdot A_{Rf}}{60 \cdot Q}\right) \cdot \frac{c_{r,u} \cdot u_A}{g} \cdot \frac{R_R}{R_a}$$

4. The static exit loss due to premature exit of the water from the cells $h_{V,a1}$:

$$h_{V,a1} = -\frac{\int_{\varphi_1}^{\varphi_2} \frac{dA_z}{d\varphi} \cdot R_a \cdot \left(1 - \sin\,\varphi\right)}{A_q} \cdot d\varphi$$

To determine this loss, the relationship $\frac{dA_z}{d\varphi}$ is determined. Two methods (graphical, numerical) were presented in the previous chapter.

5. The dynamic exit loss $h_{V,a2}$ due to the exit of the water with speed c_3:

$$h_{V,a2} = -\frac{\int_{\varphi_1}^{\varphi_2} \frac{dA}{d\varphi} \cdot \left(c_{3,u} - u_A\right) \cdot u_A \cdot d\varphi}{g \cdot A_z}$$

To determine this loss, the relationship $\frac{dA_z}{d\varphi}$ and the exit angle α_3, as well as the relative and absolute exit speed w_3 and c_3 must be determined. It can be assumed for simplification that:

$$h_{V,a2} \approx 0.8 \cdot \frac{u_A^2}{2 \cdot g}$$

6. The losses due to free hanging $h_{V,a3}$ and wading of the wheel $h_{V,a4}$. The loss due to free hanging is determined as follows:

$$h_{V,a3} = \begin{cases} H_T & \text{for} \quad H_T \leq 0\,\text{m} \\ \frac{B \cdot H_T}{g \cdot Q} \cdot u_A^3 \cdot \left(\frac{R_a - H_T/2}{R_a}\right)^3 & \text{if} \quad H_T > 0\,\text{m} \end{cases}$$

   When the wheel wades in the tailwater, a multitude of undesirable effects occur. **Wading of the wheel should be avoided up to the nominal design flow rate Q_N by discharge versus tailwater depth tests. Above Q_N, wading of the wheel can be used to limit performance.**
7. Overfilling losses (refer to Eq. (5.10)) can be avoided by proper design of the wheel.
8. Losses due to wall friction in the cells or due to water being pulled up. These losses are not taken into account. Likewise, the loss due to air resistance is neglected.
9. Mechanical losses due to friction in the bearings of the waterwheel shaft. The mechanical loss power is determined by the friction torque M_R.

$$P_{V,mech} = \frac{M_R \cdot u_A}{R_a} = \frac{M_R \cdot 2 \cdot \pi \cdot n}{60}$$

   M_R is the friction torque, which can be determined, for example, by run-out tests.

In addition to these losses, a complete system also includes mechanical losses in the gearbox and heat, re-magnetization, and fan losses in the generator.

It is advantageous to express the losses as loss heights and not as loss powers, as these provide a direct understanding of the loss in relation to the head difference H.

6.3 Efficiency

After all loss heights have been gathered, the hydraulic efficiency of the waterwheel can be determined. This results from the quotient of useful power and theoretically available power. The power which can be used is the theoretical power minus all power losses. The latter are of a hydraulic and mechanical nature, as only the transfer point after the waterwheel's bearings is important for coupling to the gearbox:

$$P_N = P_{th} - P_{V,hydr} - P_{V,mech}$$

The hydraulic efficiency of the overshot waterwheel is now determined from the individual loss heights at the inlet and outlet:

$$\eta_{WW} = \frac{P_{N,WR}}{P_{th}} = \frac{\varrho \cdot g \cdot Q \cdot \left(H - h_V\right) - P_{V,mech}}{\varrho \cdot g \cdot Q \cdot H}$$

$$\eta_{WW} = 1 - \frac{h_{V,e} + h_{V,a}}{H} - \frac{M_R \cdot u_A}{\varrho \cdot g \cdot Q \cdot H \cdot R_a} \tag{6.3}$$

where $h_{V,e} + h_{V,a}$ is the sum of the losses at the inlet and outlet. M_R is the counter-torque due to friction in the bearings.

The efficiency of the waterwheel depends on the rotational speed n of the wheel (or the peripheral speed u_A) and the flow rate Q. Therefore, efficiency calculations must always be carried out for different flow rates, head differences and rotational speeds, as shown in the examples (Chap. 8).

For the design of the gearbox and generator, the maximum power output and the maximum torque in normal operation are of interest.[3] The usable power of the hydraulic drive is:

$$P_N = \eta_{WW} \cdot P_{th}$$

where the theoretical power is:

$$P_{th} = \varrho \cdot g \cdot \left(H + \frac{c_1^2 - c_4^2}{2 \cdot g}\right) \cdot Q$$

Usually, the speed difference of the water flow in the upper and lower water is so small that one can using just the geodetic head difference H. This applies even more to large geodetic head differences which are common for overshot waterwheels.

There is the following relationship between the torque M and the power P_N delivered by the waterwheel:

$$M = \frac{P_N \cdot 60}{2 \cdot \pi \cdot n} \tag{6.4}$$

This result is also important for the design of waterwheels in order to achieve small torques.[4] To reduce the torque one must choose high rotation speed and smaller wheel radii.

Very slowly rotating waterwheels with a large radius R_a generate very high torques, which are paid for with high costs for the waterwheel shaft and for the

[3] The starting torque for waterwheels for power generation is small, as the generator is only connected to the grid when the nominal speed is reached and then provides a counter-torque. A wheel coupled to the working machines via a transmission behave differently, as it first has to overcome the friction of the transmission system and the possibly uncoupled working machines (e.g. hammer mills).

[4] This torque is the input torque of the transmission/gear system and therefore must be used for transmission dimensioning.

transmission. In addition, a low specific discharge capacity as well as an early onset of overfilling of the cells is achieved. Overshot and backshot waterwheels should run at the highest peripheral speed, which still yields a favorable efficiency η.

For the owner of a hydroelectric power plant, the plant efficiency, the electrical power, and the energetic potential are of interest.

The total plant efficiency results from the product of the efficiencies of the waterwheel, the transmission, and the generator:

$$\eta_{plant} = \eta_{WW} \cdot \eta_{trans} \cdot \eta_{gen}$$

The electrical output power of the system is obtained by multiplying the plant efficiency with the theoretically available power:

$$P_{elec} = P_{th} \cdot \eta_{plant}$$

This value is used for the dimensioning of the generator. It is also used to determinate power duration curves.

The available energetic potential (annual work of the whole plant) is:

$$W = \int_{0}^{t=365} P_{elec} \cdot dt \tag{6.5}$$

The total work of the plant is then determined by integrating the power P_{elec} over the year. Usually, the energetic potential is determined using a flow duration curve (e.g. Example 10.12). From this, the power duration line is determined. The area enclosed the power duration curve of the plant and the x-axis marked with W in Fig. 10.17 represents the annual work of the plant.

According to the German Renewable Energy Act (Erneuerbare-Energien-Gesetz) of January 1, 2012, energy supply companies are obligated to purchase the energy and compensate accordingly (depending on the installed power up to 0.127/kWh), if the plant meets the ecological requirements per water law notice. This only applies to plants that were approved after August 1, 2004. For plants that were put into operation before July 31, 2004, this compensation rate is only granted if the plant has been modernized and the water ecology has been improved according to the guidelines. Otherwise, the existing compensation of 0.0767/kWh is maintained.

It should be noted that the annual energy calculated here is further reduced in reality, as the waterwheel plant has downtime due to maintenance, ice drift, and low water times. A rough estimate of the annual work is obtained if one assumes 200 to 220 full load days. Further information on the planning and design of whole hydroelectric power plants can be found, for example, in Weiss [2], Palffy [3], Giesecke [4].

The annual work W is an important planning parameter for energy production of hydropower plants. It determines the profitability of the investment.

Fig. 6.3 Overshot waterwheel with low gross head and high discharge capacity. The inlet plate has an exit edge, which is similar to a sine curve. This results in a well-guided, variable water inflow and the impacts are reduced. In addition, the cells of the two segments of the wheel have an offset to each other. Both measures together allow a more uniform operation of the waterwheel. Photo: courtesy of Martin Impler GmbH

It is always better to use the generated mechanical or electrical energy directly, rather than feed it into the main grid. This significantly increases the profitability of the plant. Possible uses include home electricity in parallel grid operation, heat generation via a heat pump, charging of electric vehicles or traditional use through mill operation, saw frames, etc. (Fig. 6.3).

However, it should not go unmentioned that waterwheels with small rated output power are always on the borderline of profitability. The application process usually requires the same procedure and thus the same effort as for larger turbine systems. Often the technical details are clarified with a lot of effort and attention to detail. An economic examination of the system, however, does not take place. Here, publicly funded energy consulting or a professionally secured cost-benefit analysis can provide a remedy.

References

1. Bollrich G, Preißler G (1992) Technische Hydromechanik Band 1, 3rd edn. Verlag für Bauwesen, Berlin
2. Weiss P (1991) Ein Beitrag zur Planung und Projektierung von Kleinwasserkraftanlagen. PhD thesis, RWTH-Aachen, in „Mitteilungen, Institut für Wasserbau und Wasserwirtschaft, RWTH-Aachen "
3. Palffy SO (1994) Wasserkraftanlagen – Klein- und Kleinstkraftwerke, vol 322, 2nd edn. expert-Verlag Renningen-Malmsheim, Renningen-Malmsheim
4. Giesecke J, Mosonyi E, Heimerl S (2009) Wasserkraftanlagen – Planung, Bau und Betrieb, 5th edn. Springer Verlag, Berlin

Chapter 7
The Drive System

7.1 Gearbox

After the power and the optimal speed of the waterwheel can be calculated, one must consider the utilization of the water energy. The complete drive train consisting of the waterwheel, gearbox and the working machine (generator) will be described in this section.

The model of this system is shown in Fig. 7.1. A three-stage gearbox is assumed. The questions to be answered for the drive system range from the selection of the gearbox and the proper generator to the determination of the torques and dynamic processes in the drive system.

The gearbox serves to adapt the low waterwheel speed to a suitable generator or to corresponding working machines. For a waterwheel for power generation, gear ratios of $i = 42 \ldots 340$ should be realized.[1]

One problem to be dealt with is the very large torques $M = \frac{P_N}{\omega}$. Such large values occur because the waterwheels have a low rotational speed. Gearboxes are therefore a decisive financial factor in waterwheel construction. The cost of a gearbox can be similar to that of the waterwheel itself.

For an ideal gearbox (no slip, no friction) the relationships given in Table 7.1 apply for the point of engagement, since the same paths are covered at the point of engagement. Thus the speeds are equal, and for the force at the transmission point, the principle of actio equals reactio applies. With the relationships from Table 7.1 one obtains the gear ratio:

$$i = \frac{\omega_1}{\omega_2} = \frac{n_1}{n_2} = \frac{M_1}{M_2} = \frac{R_1}{R_2} \tag{7.1}$$

[1] Corresponding to generator speeds of approximately $510 \ldots 1020\ \text{min}^{-1}$.

D. M. Nuernbergk, *Overshot Waterwheels*, Springer Tracts in Mechanical Engineering, https://doi.org/10.1007/978-3-031-95385-9_7

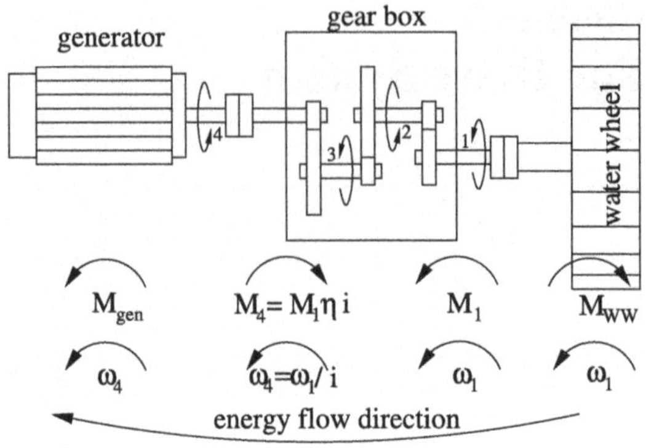

Fig. 7.1 Energy flow diagram and torque conversion model for a rigidly coupled arrangement waterwheel—gearbox—electric machine. At the interfaces (couplings), $M_1 - M_{WW} = 0$ and $M_{gen} - M_1 \cdot \eta_{gear} \cdot i = 0$ apply in equilibrium

Table 7.1 Physical relationships at a transmission point between two rotating gearbox wheels

Physical quantity	Gear wheel 1		Gear wheel 2
Path $s =$	$R_{a1} \cdot \varphi_1$	$=$	$R_{a2} \cdot \varphi_2$
Circumferential speed $u =$	$R_{a1} \cdot \omega_1$	$=$	$R_{a2} \cdot \omega_2$
Magnitude of force $F =$	M_1 / R_{a1}	$=$	M_2 / R_{a2}
Power P	$M_1 \cdot \omega_1$		$M_2 \cdot \omega_2$
Kinetic energy W_{kin}	$J_1 \cdot \frac{\omega_1^2}{2}$		$J_2 \cdot \frac{\omega_2^2}{2}$

It is important in a transmission system that the torque and the speed is increased. The transmission has the function of speed and torque adjustment. For a multi-stage transmission arrangement, one obtains (Fig. 7.1):

$$i = i_{12} \cdot i_{23} \cdot i_{34}$$

for the total gear ratio. From the effective torques, the magnitude of the force on the teeth at the point of engagement can be calculated:

$$F = \frac{M_1}{R_1} = \frac{M_2}{R_2}.$$

Now it also becomes clear why the radius of the first gear wheel had to be chosen very large in the past (wooden teeth in spur gears). The radius was chosen to keep the tooth pressure small and not only to realize of the large gear ratio i.

For an ideal transmission $P = M_1 \cdot \omega_1 = M_2 \cdot \omega_2 = M_3 \cdot \omega_3 = M_4 \cdot \omega_4$, torques are converted as follows:

$$M_4 = M_1 \cdot \frac{\omega_1}{\omega_4} = M_1 \cdot i_{12} \cdot i_{23} \cdot i_{34} = M_1 \cdot i$$

When converting the torques, the direction of energy flow must be taken into account, as losses occur in the transmission element (gearbox). It holds that:

$$M_4 = M_1 \cdot \eta_{gear} \cdot i$$

M_4 is the torque of the waterwheel referred to the generator shaft and i is the total transmission ratio. In principle, the other direction of energy flow is also conceivable. The electric machine in motor operation then drives the waterwheel. This case is possible, for example, when using asynchronous generators with short-circuit rotors. It is undesirable and must be suppressed.

Another important quantity is the total moment of inertia J of the drive system. The moment of inertia is a measure of the inertia that a rotating system opposes to an acceleration or deceleration of the rotational movement. It is given by:

$$J \cdot \frac{\omega_1^2}{2} = J_1 \cdot \frac{\omega_1^2}{2} + J_2 \cdot \frac{\omega_2^2}{2} + J_3 \cdot \frac{\omega_3^2}{2} + J_4 \cdot \frac{\omega_4^2}{2}$$

Here, $J_1 \ldots J_4$ are the moments of inertia for all parts arranged on the shafts and the shafts itself. The values can be calculated according to the formulas according to, e.g., Dubbel [1]. Substituting the corresponding gear ratios for the angular velocities, the following applies:

$$\omega_1 = i_{12} \cdot \omega_2 \quad \text{or} \quad n_1 = i_{12} \cdot n_2$$

$$\omega_2 = i_{23} \cdot \omega_3 \quad \text{or} \quad n_2 = i_{23} \cdot n_3$$

$$\omega_3 = i_{34} \cdot \omega_4 \quad \text{or} \quad n_3 = i_{34} \cdot n_4$$

Inserting this value into the above formula, one obtains:

$$J = J_1 + J_2 \cdot \left(\frac{1}{i_{12}}\right)^2 + J_3 \cdot \left(\frac{1}{i_{12} \cdot i_{23}}\right)^2 + J_4 \cdot \left(\frac{1}{i_{12} \cdot i_{23} \cdot i_{34}}\right)^2$$

as the total moment of inertia referred to the waterwheel shaft. The total moment of inertia is increased compared to the moment of inertia of the individual shaft. Analogously, the total moment of inertia referred to the generator shaft is obtained:

$$J = J_1 \cdot (i_{12} \cdot i_{23} \cdot i_{34})^2 + J_2 \cdot (i_{23} \cdot i_{34})^2 + J_3 \cdot (i_{34})^2 + J_4$$

The equation of motion of the drive system on the generator side is then:

$$J \cdot \frac{d\omega}{dt} = M_1 \cdot \eta_{gear} \cdot i - M_{gen} \tag{7.2}$$

M_{Gen} is the counter torque of the generator and J is the total moment of inertia referred to the generator shaft. In all non-uniform movements (starting of the waterwheel, load drop with speed increase to the maximum speed, imbalance of the waterwheel), the acceleration torques shape the motion sequence of the generator.

Table 7.2 Characteristics of selected gearboxes (orientation values—exact characteristics are provided by the gearbox manufacturer)

Type of gearbox	Gear ratio $i = \frac{n_{an}}{n_{ab}}$	Gearbox efficiency η_{gear}
Spur gear	Up to 8 single-stage	0.99...0.96
	6...45 two-stage	0.97...0.91
	30...250 three-stage	0.95...0.85
Worm gear	Up to 60	0.70...0.50 single-entry
		0.80...0.70 double-entry
Belt drive	up to 8	0.97...0.94
Chain drive	up to 6	0.98...0.97
Friction drive	up to 6	0.98...0.95

To avoid, for example, flicker phenomena in the electrical grid, waterwheels should be as well balanced as possible.

For instance, during standstill times, a waterwheel with wooden cells is affected by imbalance when the wheel dips into the underwater or is affected by a leakage water flow. The waterwheel should be regularly turned a quarter of the wheel circumference to avoid such an imbalance. This measure results in a uniform soaking of the wooden cells.

Table 7.2 lists some characteristics of gearboxes, but these can only be guideline values, as the gearbox efficiency depends on the load and the speed (see Chap. 9, Fig. 10.5 and the corresponding text there). Waterwheels need a three- to four-stage transmission.

Often the first stages are spur gears $i = 4\ldots6$ and the last stage is a belt drive $i = 3\ldots6$ connected to the generator. This allows the speed of the waterwheel to be adjusted afterwards by simply changing the pulleys. For gearboxes with spur gears, only gearboxes from heavy machinery or mining are suitable due to the large torques which have to be handled. These are now available in very compact design.

Recently, there have been attempts to improve the bearings and gear box system. Concepts such as flying bearing and planetary gear transmission (Turas waterwheel [2]) were applied. Furthermore, a concept was applied which avoids a waterwheel bearing on one side by using the gear box bearing [3]. However, it should be noted that when operating waterwheels, a permanent, cyclic load occurs and very long service lives of the plant are to be expected. Under such conditions, a one-sided bearing can be problematic.

In addition, one must expect extreme situations, such as floods and ice formation, as shown in Figs. 7.2 and 7.3.

An overview can be found in a handbook for mechanical engineering like e.g. Dubbel [1] for the dimensioning and design of shafts, bearings and gears. Detailed drawings of hubs (rosettes), shafts and bearings can be found in the atlases of the books of Henne [4], Müller [5] and Bach [6].

Fig. 7.2 Warendorf waterwheel (Turas waterwheel) under flood conditions. The waterwheel is almost submerged up to the wheel shaft in the underwater

Fig. 7.3 Warendorf waterwheel under flood conditions. The cells lift water upwards in the direction of rotation, as the ventilation of the cells cannot occur quickly enough

Example 7.1. ***(Calculation of a Transmission)*** **A waterwheel with a diameter of $R_a =$ 2.5 m rotates at a constant speed of $n = 5.0\,\text{min}^{-1}$. The maximum useful power output of the wheel is 15 kW. A 15 kW three-phase asynchronous motor is to be used as a generator with a synchronous speed of 1000 min^{-1}. The transmission consists of a three-stage spur gear ($\eta = 0.99$ per stage) and a flat belt drive ($\eta = 0.98$), which couples the gear box and generator. Calculate the necessary gear ratio i, the torques on the waterwheel and generator side as well as the transmission losses.**

Table 7.3 Manufacturer's specifications for speeds, efficiencies at full load and 3/4 load, the power factors at full load and 3/4 load, the rated currents, the torques, the moments of inertia and the mass at different powers of 6-pole asynchronous machines for $f = 50$ Hz, rated speed 1000 min^{-1}

P_N [kW]	n $\left[\frac{1}{min}\right]$	$\eta_1/\eta_{3/4}$ [%]	$\cos\varphi_1/\cos\varphi_{3/4}$ [1]	I_N [A]	$\frac{I_S}{I_N}$ [1]	M_N [Nm]	$\frac{M_S}{M_N}$ [1]	$\frac{M_{max}}{M_N}$ [1]	J [kgm^2]	m [kg]
2.2	945	83.6/84.2	0.74/0.67	5.20	6.0	22.2	2.4	2.9	0.018	33
3	965	86.7/87.1	0.79/0.73	6.40	6.9	29.7	2.4	2.9	0.038	46
4	960	87.4/88.2	0.80/0.74	8.30	7.0	39.7	2.9	2.9	0.045	54
5.5	960	87.0/87.6	0.77/0.70	12.0	6.9	54.7	2.4	2.8	0.049	59
7.5	975	90.7/91.1	0.77/0.71	15.6	7.8	74.0	2.2	3.5	0.107	102
11	970	90.9/91.5	0.78/0.72	23.0	7.6	109	2.8	3.4	0.127	117
15	980	92.2/92.4	0.77/0.71	31.0	8.3	147	2.5	3.9	0.237	160
18.5	985	92.4/92.4	0.81/0.76	36.0	8.0	179	4.3	3.5	0.43	185
22	985	92.9/93.0	0.82/0.78	42.0	8.1	214	4.3	3.2	0.49	200
30	985	93.6/93.8	0.83/0.80	56.0	7.5	291	3.8	2.9	0.75	252

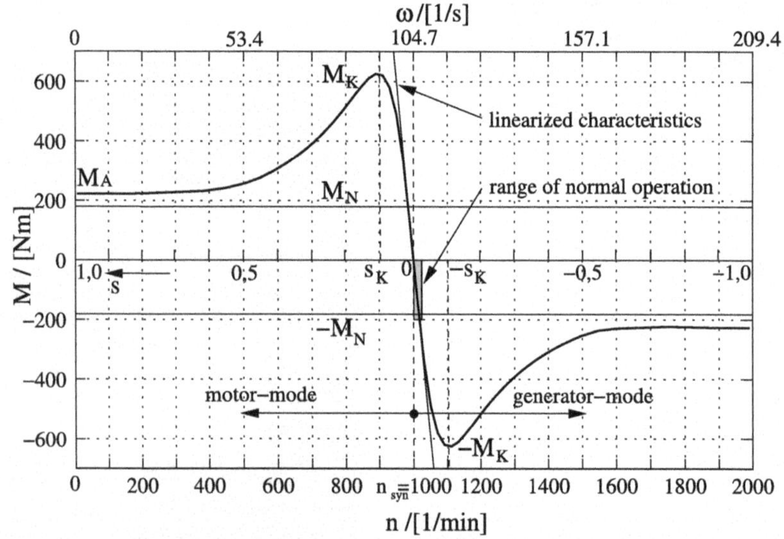

Fig. 7.4 Torque-speed characteristic (or slip characteristic) of an asynchronous machine. The nominal power was $P_N = 18.5$ kW, $M_{max}/M_N = 3.5$, $M_N = 179$ Nm

From Table 7.3, it can be seen that a generator speed of approximately 1020 min^{-1} (same nominal torque as at 980 min^{-1}) should be achieved to reach the counter torque for the waterwheel rotating at 5.0 min^{-1} (see also Fig. 7.4):

$$i = \frac{n_1}{n_5} = \frac{5.0}{1020} = 0.0049$$

$$1 : i = 1 : 204$$

The torque delivered by the waterwheel is (indices see Fig. 7.1):

$$M_1 = \frac{P_N \cdot 60}{2 \cdot \pi \cdot n_1} = \frac{15000 \cdot 60}{2 \cdot \pi \cdot 5} = 28650\,\text{Nm}$$

On the generator side, the torque delivered by the transmission is:

$$M_5 = \eta_{gear} \cdot i \cdot M_1 = 0.99^3 \cdot 0.98 \cdot 0.0049 \cdot 28650 = 133.5\,\text{Nm}$$

This torque is with the generator counter torque M_{gen} in balance. The mechanical power $P_{mech,gen}$ available on the generator side is:

$$P_{mech} = M_4 \cdot \omega_4 = 133.5 \cdot \frac{2 \cdot \pi \cdot 1020}{60} = 14260\,\text{W}$$

The gearbox has a power loss of $P_V = 15000\,\text{W} - 14260\,\text{W} = 740\,\text{W}$ at full load, which is almost completely converted into heat. An increase in the speed of the waterwheel to $n = 5.5\,\text{min}^{-1}$ would have reduced the torque to $M_1 = 26040\,\text{Nm}$. The gear ratio would then be $1 : i = 1 : 185.5$. The gearbox efficiency and the torque on the generator side remain approximately the same. High (even slightly higher) speeds of the waterwheels are absolutely desirable in order to keep the torques small. With the specifications gear ratio, input torque M_1, output torque M_5 and power loss, a gearbox can now be selected from conventional gearbox catalogs.

7.2 Generators

The usual operating modes for small hydroelectric plants are:

1. Feeding into the grid, grid-connected operation
2. Parallel grid operation, and
3. Island operation.

The generator should be chosen depending on the mode of operation and the costs. As operating speeds of generators, the speeds $n = 510\ \text{min}^{-1}$, $n = 610\ \text{min}^{-1}$, $n = 770\ \text{min}^{-1}$, $n = 1020\ \text{min}^{-1}$ are often used in small hydroelectric power plants [7]. The operating speed is thus slightly above the synchronous speed of the generators. As a rule, the higher the synchronous speed or the nominal power of a generator is, the higher is the generator efficiency. In addition, the generator with higher synchronous speed is smaller and more cost-effective (fewer pole pairs and thus smaller size of the electrical machines).

Asynchronous generators,[2] synchronous machines and synchronous machines with permanent magnets [8] are used as generators. For grid-connected operation and parallel grid operation, three-phase asynchronous machines with a short-circuit rotor are used. These generators are affordable and no expensive synchronization device is necessary, as the generator is switched on at synchronous speed. This

[2] The asynchronous motor is simply used in generator operation.

type of generator will be examined and calculated in more detail in this section in conjunction with a waterwheel-gear arrangement.

Other machines can be designed in a similar way if the torque-speed characteristic is known. An overview of rotating machines, types of machines, power plates, protection types, connection assignments and designs can be found in Friedrich [9].

7.2.1 Asynchronous Generators

Asynchronous machines (ASM) with short-circuit rotor are often used in hydroelectric power plants, as they have the following properties and advantages:

1. robust construction and low maintenance,
2. operation at only one speed rigidly on the grid (in the absence of speed switching),
3. operation directly on the grid (reactive power from the grid),
4. variable speed operation with a frequency converter possible,
5. basic series of motors for universal use from a few watts to 10 MW power are available (the power gradation is approximately 1 : 1.25),
6. there is a comprehensive range of protection degrees, protection types and designs,
7. at 400 V operating voltage, synchronous speeds (ideal idle speed) of $n_{syn} = 500, 600, 750, 1000, 1500, 3000\,\text{min}^{-1}$ are available and
8. the switching between star and delta connection allows the choice of voltage between $U = 230\,\text{V}$ and $U = 400\,\text{V}$.

In Tables 7.3 and 7.4 the characteristic values of high-efficiency asynchronous machines with a synchronous speed of $n_{syn} = 1000$ and $750\ \text{min}^{-1}$ are listed.

The mode of operation is based on the rotating field in the air gap between the stator and rotor (runner) of the machine. This rotating field induces voltages in the stator and rotor windings. The rotor current forms the torque with the rotating field and is transferred like in a transformer to the stator side.

Under load, a slip occurs between the rotating field and the rotor. The larger the slip is, the larger is the torque. The rotor is very simple in construction. It consists of bars that are short-circuited at the ends of the rotor by rings (squirrel-cage rotor). The absence of brushes makes this electrical machine particularly low-maintenance.

7.2.2 Operating Characteristics of the Asynchronous Generator

The slip is the deviation of the machine speed n from the synchronous speed n_{syn}, relative to the synchronous speed. It is defined as:

Table 7.4 Manufacturer's specifications for speeds, efficiencies at full load and 3/4 load, the power factors at full load and 3/4 load, the nominal currents, the torques, the moments of inertia and the mass at different powers of 8-pole asynchronous machines for $f = 50\,\text{Hz}$, nominal speed $750\ \text{min}^{-1}$

P_N [kW]	n $\left[\frac{1}{\text{min}}\right]$	$\eta_1/\eta_{3/4}$ [%]	$\cos\varphi_1/\cos\varphi_{3/4}$ [1]	I_N [A]	$\frac{I_S}{I_N}$ [1]	M_N [Nm]	$\frac{M_S}{M_N}$ [1]	$\frac{M_{max}}{M_N}$ [1]	J [kgm^2]	m [kg]
2.2	725	83.7/83.3	0.66/0.57	5.9	5.6	29.1	2.1	3.0	0.045	53
3	715	82.8/83.1	0.69/0.61	7.7	5.2	40.1	2.3	2.6	0.049	59
4	720	86.3/87.1	0.69/0.62	9.8	5.2	53.0	2.2	2.6	0.091	88
5.5	725	86.9/87.3	0.66/0.58	14.1	6.0	73.0	2.3	3.1	0.131	118
7.5	715	87.0/87.7	0.70/0.63	18.2	5.4	100	2.1	2.7	0.131	118
15	735	92.1/92.5	0.81/0.77	29	7.8	195	3.1	3.4	0.540	200
18.5	735	92.4/92.8	0.76/0.70	38	6.5	240	2.6	3.0	0.680	225
22	735	92.5/92.8	0.75/0.69	46	7.1	286	3.0	3.4	0.800	255
30	740	93.6/94.0	0.78/0.73	60	7.6	388	2.8	3.3	1.520	320

$$s = \frac{\omega_{syn} - \omega}{\omega_{syn}} = \frac{n_{syn} - n}{n_{syn}} = 1 - \frac{n}{n_{syn}}$$

The synchronous angular speed is calculated as

$$\omega_{syn} = \frac{2 \cdot \pi \cdot f}{p}$$

and the synchronous rotational speed is

$$n_{syn} = \frac{\omega_{syn} \cdot 60}{2 \cdot \pi} = \frac{f \cdot 60}{p}$$

The term "synchronous speed" (also ideal idle speed) indicates that the rotor of the asynchronous machine rotates synchronously with the net frequency f/p. In these equations, f is the grid frequency[3] and p is the number of pole pairs.

Two points are important for the operating characteristics: for $s = 0$ there is idle operation and for $s = 1$ the machine is at a standstill. The torque characteristic in motor and generator operation is of particular interest when coupling with a waterwheel gear arrangement. The torque of the asynchronous machine is given by the equation:

$$M = M_K \cdot \frac{2 \cdot s \cdot s_K}{s^2 + s_K^2}$$

[3] This frequency is $f = 50\,\text{Hz}$ in Europe and $f = 60\,\text{Hz}$ in North America.

M_K is the tipping torque and s_K is the slip at the tipping point. The characteristic curve of the asynchronous machine (abbreviated ASM) shown in Fig. 7.4 has further distinctive points: the tipping point at s_K, M_K, where the ASM delivers its maximum torque, and the point of the starting torque M_A at standstill.

The ratio between the tipping torque and the rated torque is $M_K/M_N = 2.5 \ldots 4$ (see Tables 7.3 and 7.4). The starting torque is only of interest in motor operation, as the starting torque must be greater than the rated torque. Otherwise, the motor does not start.

In Fig. 7.4 the torque characteristic of an ASM is shown. The values for this characteristic were taken from Table 7.3 for an 8-pole ASM with a rated power of $P_N = 18.5$ kW. For negative slip values $s < 0$ (or speeds which are higher than the synchronous speed) the ASM acts as a generator. The operating range of the asynchronous machine as a generator is shown hatched.[4]

The equation for the torque can be linearized around the point of synchronous speed. The following equation is a good approximation:

$$M = M_K \cdot \frac{2 \cdot s}{s_K} = \frac{2 \cdot M_K}{s_K} \cdot \left(1 - \frac{n}{n_{syn}}\right)$$

In Fig. 7.4, this linearized characteristic is drawn as a straight line. This can be easily determined with the values from Table 7.3, as for the rated power, the rated torque M_N and the machine speed are given. The second point of the characteristic is known because for synchronous speed the torque equals $M = 0$.

The speed increases slightly with increasing power output of the waterwheel. For an adapted speed control, one would wish for a much flatter torque characteristic of the ASM, like the graph shown in Fig. 7.4.

From the balance of torques between the waterwheel torque referred to the generator shaft and the counter torque of the generator, one can now determine the rotational speed. This speed depends on the theoretical power (water flow, head).

If there is no balance of these moments, the speed of the system must change according to the equation of motion. As prerequisites, a constant waterwheel and gear efficiency are assumed and as a load on the generator the public grid is assumed.[5]

The waterwheel torque M_{WW} converted to the generator shaft must be equal to the generator torque M_{en}:

$$M_1 \cdot \eta_{gear} \cdot i - M_{Gen} = 0$$

$$\frac{60 \cdot P_{th} \cdot \eta_{WW}}{2 \cdot \pi \cdot n_1} \cdot \eta_{gear} \cdot i + \frac{2 \cdot M_K}{s_K} \cdot \left(1 - \frac{n}{n_{syn}}\right) = 0$$

[4] Another representation of the characteristics of an ASM can be found in Fig. 10.21.

[5] In island operation, speed changes occur due to load fluctuations. These must be compensated by a load controller [10].

Between the generator speed and the waterwheel speed, $n_1 = n \cdot i$ applies. Then, one obtains the following quadratic equation for the speed at the generator:

$$n^2 - n_{syn} \cdot n - \frac{60 \cdot P_{th} \cdot \eta_{WW} \cdot \eta_{gear} \cdot n_{syn}}{2 \cdot \pi \cdot \frac{2 \cdot M_K}{s_K}} = 0$$

The solution to the quadratic equation is:

$$n = \frac{n_{syn}}{2} \pm \sqrt{\frac{n_{syn}^2}{4} + \frac{60 \cdot P_{th} \cdot \eta_{WW} \cdot \eta_{gear} \cdot n_{syn}}{2 \cdot \pi \cdot \frac{2 \cdot M_K}{s_K}}} \tag{7.3}$$

where the solution with the positive sign is to be used for generator operation. The negative sign is used for motor operation. The solution shows an almost linear relationship between the generator speed and the power of the waterwheel.

The corresponding torque of the generator M can now be calculated with the linearized characteristic curve for a given rotational speed and power. The generator delivers the power P_{el} to the electrical grid when the corresponding torque M is effective on the machine. It holds that:

$$P_{elec} = M \cdot \omega \cdot \eta_{gen}$$

This describes the interface between mechanical and electrical quantities, and all quantities in the drive system are determined now, since the grid frequency is a constant and thus $n_{syn} = const$ is given by the generator specification.

The ASM can be connected in star or delta configuration, depending on whether all six strand ends are led out of the machine. Measuring the conductor currents I and conductor voltages U in the supply lines, the following power formulas apply for both star and delta configuration:

$$P_{elec} = \sqrt{3} \cdot U \cdot I \cdot \cos\varphi$$
$$Q_{elec} = \sqrt{3} \cdot U \cdot I \cdot \sin\varphi$$
$$S_{elec} = \sqrt{3} \cdot U \cdot I$$

$\cos\varphi$ is the power factor, P_{el} the active-, Q_{el} the reactive- and S_{el} the apparent power. U is the line-to-line voltage and I is the line current.

The ASM in generator operation draws inductive reactive power from the grid. This arises because inductances and capacitances in the AC grid store and release energy. The reactive power oscillates in the AC grid between the generator and the consumer. The reactive power for the ASM is:

$$Q_{el} = P_{el} \cdot \frac{\sqrt{1 - \cos^2\varphi}}{\cos\varphi}$$

To distinguish from the active power (unit W = Watt), the reactive power is given in the unit $kvar$ (volt-ampere reactive). The reactive power component is undesirable. A poor power factor $\cos\varphi$ thus burdens the public distribution grid. The resulting reactive power component can be compensated by switching on capacitors. The size of the compensation capacitor can be estimated as follows:

$$Q_C = P_{el} \cdot (\tan\varphi - \tan\varphi_1)$$

$$C = \frac{Q_C}{U^2 \cdot \omega}$$

$\cos\varphi$ is the power factor and φ the angle before compensation and $\cos\varphi_1$ the power factor and φ_1 the angle after compensation. Q_C is the compensation reactive power, C the compensation capacity, ω the angular frequency of the fixed network, U the voltage and P_{el} the electrical power. By switching on capacitors, the power factor of the arrangement generator/capacitors and thus the network quality is improved. The capacitor power Q_C is 35 to 50% of the rated power of the three-phase machines [9].

For the design of an ASM as a generator, the following applies:

1. The generator be designed with appropriate power close to the nominal power (power-appropriate design), as at low design power the power factor $\cos\varphi$ and the efficiency η decrease. A generator that is too large is expensive, has a poor efficiency, requires more space, and the cables for connection to the grid are also oversized. This increases the system costs.
2. The generator may be loaded with a power that is $5 \ldots 10\%$ greater than the nominal power. However, overloading reduces the lifespan of the electric machine and causes significant heating.
3. As the synchronous speed decreases, the efficiency and the power factor decrease, and the price and the mass of the machine increase. A cost calculation must be carried out to determine what an increase in gear ratio costs compared to a generator with higher synchronous speed. The annual energy yield must be taken into account in this calculation.
4. The reactive power component of the ASM must be compensated by compensation capacitors.

Variable speed operation of the ASM is also possible through the following options:

1. Pole switching,
2. Frequency control by frequency converters, which influences the synchronous speed of the ASM,
3. Stator voltage control,
4. Resistance control (lossy).

In hydropower plants with waterwheels, frequency control with frequency converters and pole switching have been used for speed control [3]. The decision to use a frequency converter ultimately depends on a cost-benefit analysis, since the conversion is also associated with losses (4–6%). Speed control with a pole-switchable generator is much cheaper. This allows for operation at two different rotational speeds.

For further reading, refer to references on electrical drive technology [11–13]. For overshot waterwheels, speed control is not needed under properly chosen operating conditions.

Example 7.2. ***(Design of a Three-Phase Asynchronous Generator)*** **Determine the speed of a three-phase asynchronous generator** ($P = 7.5\,\text{kW}$, $n_{syn} = 750\,\text{min}^{-1}$), **when the mechanical power output of the waterwheel, which is coupled via a gearbox** ($i = 1 : 171$, $\eta_{gear} = 0.95$), **changes from** $P_N = P_{th} \cdot \eta_{WW} = 3$–$7\,\text{kW}$. **The electrical active and reactive power must be calculated.**

A rotational speed of $715\,\text{min}^{-1}$ in motor mode can be found for a nominal power of $P_N = 7.5\,\text{kW}$ in Table 7.4. The nominal torque is $M_N = 100\,\text{Nm}$. From these details, the previously unknown value $2 \cdot M_k / s_k$ of the linearized characteristic curve can be determined:

$$\frac{2 \cdot M_k}{s_k} = \frac{M}{s} = \frac{M}{1 - \frac{n_n}{n_{syn}}}$$

The slip is $s = 1 - \frac{n_n}{n_{syn}} = 1 - \frac{715\,\text{min}^{-1}}{750\,\text{min}^{-1}} = 0.04667$. Thus, this expression equals:

$$\frac{2 \cdot M_k}{s_k} = \frac{100}{0.04667} = 2143\,\text{Nm}$$

With Eq. (7.3), for a power of 7 kW, the resulting rotational speed is:

$$n = \frac{n_{syn}}{2} + \sqrt{\frac{n_{syn}^2}{4} + \frac{60 \cdot P_N \cdot \eta_{gear} \cdot n_{syn}}{2 \cdot \pi \cdot \frac{2 \cdot M_k}{s_k}}}$$

$$n = \frac{750}{2} + \sqrt{\frac{750^2}{4} + \frac{60 \cdot 7000 \cdot 0.95 \cdot 750}{2 \cdot \pi \cdot 2143}}$$

$$= 778.5\,\text{min}^{-1}$$

The speeds of the generator and the waterwheel are then at different mechanical power output P_N of the waterwheel:

$$P_N = 3000\,\text{W} \quad n_{gen} = 762.5\,\text{min}^{-1} \quad n_{WW} = n_{gen} \cdot i = 4.46\,\text{min}^{-1}$$

$$P_N = 7000\,\text{W} \quad n_{gen} = 778.5\,\text{min}^{-1} \quad n_{WW} = n_{gen} \cdot i = 4.55\,\text{min}^{-1}$$

The change in generator and waterwheel rotational speed at different power levels is small. The active electrical power of the generator is (the efficiency of the generator is taken from Table 7.4):

$$P_{elec} = P_N \cdot \eta_{gear} \cdot \eta_{gen} = 7000 \cdot 0.95 \cdot 0.87 \quad = 5785\,\text{W}$$

$$P_{elec} = P_N \cdot \eta_{gear} \cdot \eta_{gen} = 3000 \cdot 0.95 \cdot 0.877 = 2500\,\text{W}$$

and the reactive power is respectively:

$$Q_{elec} = P_{elec} \cdot \frac{\sqrt{1-\cos^2\varphi}}{\cos\varphi} = 5785 \cdot \frac{\sqrt{1-0.7^2}}{0.7} = 5902\,\text{Var}$$

$$Q_{elec} = P_{elec} \cdot \frac{\sqrt{1-\cos^2\varphi}}{\cos\varphi} = 2500 \cdot \frac{\sqrt{1-0.63^2}}{0.63} = 3082\,\text{Var}$$

The absorbed reactive power is inductive in the ASM.

7.2.3 Island Operation with Asynchronous Machines

The advantages of the ASM compared to permanent- or current-excited synchronous machines lie in the price. Maher [14] shows a way how three-phase asynchronous machines up to 5 kW can be connected as a generator for small Pelton turbine hydroelectric power plants in island operation. The basic principle of exciting an ASM with a C-2C network is shown in Fig. 7.5.

A single-phase alternating current is generated with this method. The necessary capacity values C for the network are calculated as follows [14]:

$$C = 0.95 \cdot \frac{I_N}{U_K \cdot 4 \cdot \pi \cdot f}$$

I_N is the nominal current from one strand, U_K is the terminal voltage of the generator and f is the frequency of the generator. For this mode of operation, a load regulator is necessary, which switches on or off a variably adjustable basic load (e.g. various heating coils, electric heating) in case of load changes and thus stabilizes the frequency and voltage. Furthermore, the network must be adequately protected against touch, short circuit, etc.

Protective regulations can be found, for example, in Friedrich [9], and detailed information on the installation of such an autonomous network can be found in Maher [14]. The manufacturer addresses of load controllers are also mentioned in

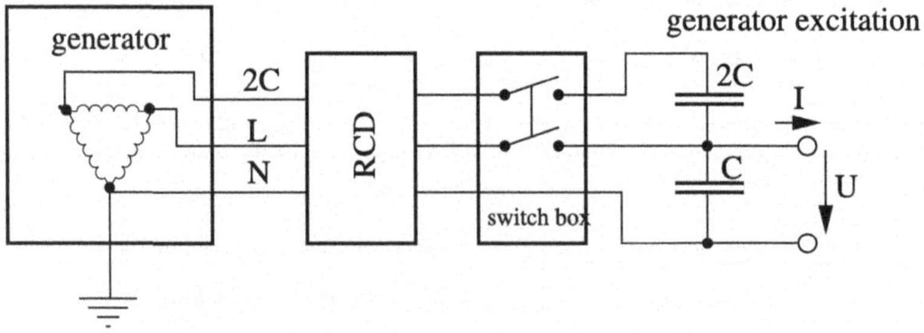

Fig. 7.5 Excitation of an asynchronous machine with a C-2C network

Maher [14]. The implementation of such a network should only be carried out by trained professionals.

7.2.4 Star-Delta Switching

There is a simple way to increase the efficiency of an existing system without having to change the three-phase asynchronous generator. This is particularly true when the system is very often operated in the lower partial load range, as is typical for small hydroelectric plants. In the partial load range, the efficiency of the machine set and the generator is usually poor.

By simply switching a three-phase asynchronous generator from delta to star configuration (Fig. 7.6), a significantly improved efficiency and a better power factor of the generator can be achieved in the lower power range (see Fig. 7.7). This was tested by the author at a hydroelectric plant with a Zuppinger waterwheel in Thuringia [15].

One result of the investigation can be found for a 22 kW generator in Fig. 7.7. From this figure, the following relationships can be seen: The efficiency in star configuration is better at low mechanical input powers than in delta configuration. At high powers, the delta configuration is advantageous. The intersection of the efficiency curves (corresponding to the switching point) is at about half of the mechanical nominal power $P_{mech} = P_{elec}/\eta$. The reason for the better efficiency of the star configuration in the partial load range are the lower ohmic losses due to lower currents in the machine.

Another very important advantage of the star configuration is the significantly better power factor of the generator. Even at a quarter of the nominal power, a

Fig. 7.6 Connections of a 22 kW generator, wired in star configuration

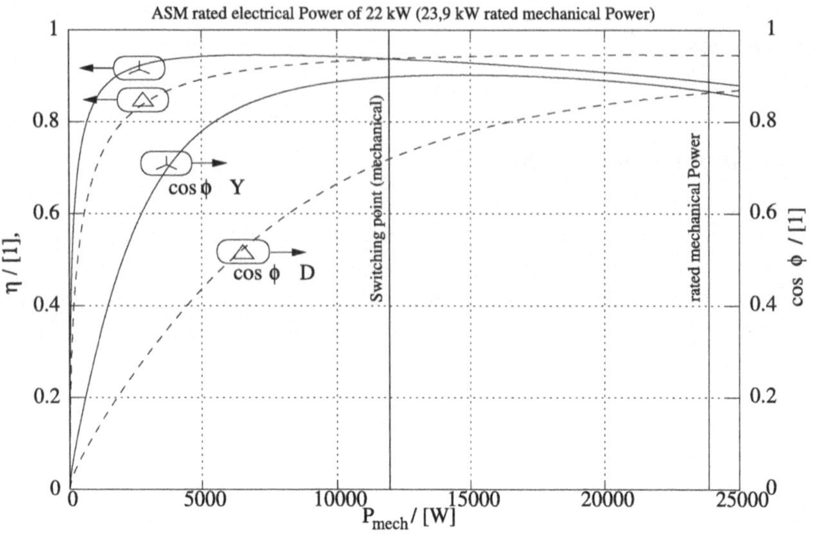

Fig. 7.7 Efficiency η and power factor $\cos\varphi$ for a three-phase asynchronous generator $P_{elec} =$ 22 kW in star and delta configuration. The results were determined using Spice simulations for a model of a three phase ASM

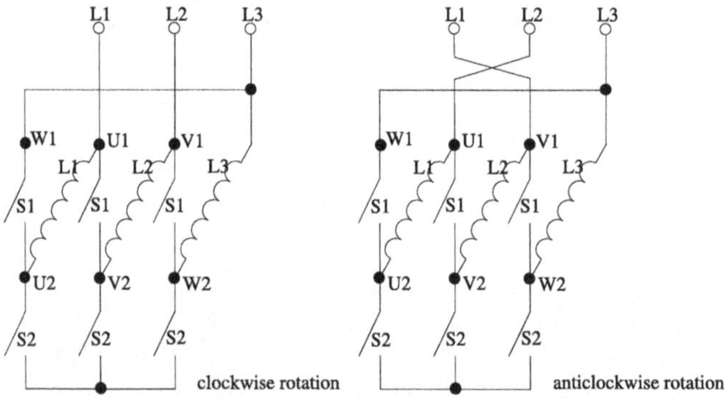

Fig. 7.8 Star-delta switching of the coils L1, L2, L3 of a three-phase asynchronous machine for clockwise and counterclockwise rotation

power factor approximately equal to the nominal value is achieved. The reactive currents in the partial load range are thus much lower, which in turn reduces the losses in the generator. However, it should be noted that with reduced reactive currents, the compensation of the plant must be adjusted accordingly by switching the compensation capacitors from star to delta connection as well.

It is therefore worthwhile to switch three-phase asynchronous generators below half the nominal power from delta to star because of the better efficiency of the generator. The circuit diagram for the switching can be found in Fig. 7.8. The efficiency of hydroelectric power plants will increase by a few percent in the partial load range with this measure. This behavior could also be experimentally demonstrated on a Zuppinger waterwheel [15].

The switching of the system must occur when the generator is at a standstill to avoid a jerk in the system, as the generator changes its speed during the switching process. A simple star-delta switch is sufficient for this purpose.

This results in a change of the rotational speed [15], which reads:

$$n_Y = \frac{n_{syn}}{2} \pm \sqrt{\frac{n_{syn}^2}{4} + 3 \cdot \left(n_\Delta^2 - n_\Delta \cdot n_{syn}\right)} \tag{7.4}$$

and

$$n_\Delta = \frac{n_{syn}}{2} \pm \sqrt{\frac{n_{syn}^2}{4} + \frac{\left(n_Y^2 - n_Y \cdot n_{syn}\right)}{3}} \tag{7.5}$$

where the positive sign in both equations applies to the generator operation and the negative sign to the motor operation. When switching from delta to star configuration in generator operation, the speed increases slightly and conversely, it decreases when switching from star to delta. For example, for a 22 kW generator with $n_{syn} = 750\ \text{min}^{-1}$ in delta configuration, the full-load speed is $n_\Delta = 765\ \text{min}^{-1}$, whereas in star configuration, the speed is $n_Y = 793.4\ \text{min}^{-1}$.

When the coils of the generator are switched from delta to star, the capacitors of the reactive power compensation must also be switched from delta to star. This reduces the capacitive reactive power of the capacitors to one third $Q_\Delta/3 = Q_Y$ and thus optimally adapts it to the lower inductive reactive power of the generator switched to star configuration. By switching the capacitors, overcompensation of the system is avoided. The calculation of the compensation scheme must be left to a specialist or should be taken from the manufacturer information of the generator.

For asynchronous machines up to 20 kW, the compensation capacitors can be connected in parallel directly to the winding coils of the asynchronous machine.

7.2.5 Synchronous Generators

The construction of the stator in the synchronous machine is the same as in the asynchronous machine. In contrast to the ASM, the rotor is excited with a winding via the excitation current that can be fed in by brushes from the outside, or permanent magnets are used. There is no slip, as the rotating field and the rotor run synchronously.

The three-phase synchronous machine has the following properties in generator operation:

1. The terminal voltage depends on the speed and the load conditions.
2. The synchronous machine falls out of synchronous speed under overload conditions.

3. The frequency depends on the speed and the number of pole pairs.
4. The machines are more expensive than ASM and require more maintenance, as the brushes are subject to wear.

Synchronous machines are used in island operation to generate the power supply of isolated or rural areas. In large power plants, as well as in emergency power units, or in reactive power machines, synchronous machines are used. The voltage and power curve depends on the type of load. With resistive and inductive load, the voltage drops. If the load is predominantly capacitive, the terminal voltage increases.

The terminal voltage can be kept constant by controlling the excitation current. With predominantly resistive and inductive load, the excitation current must be increased. Conversely, the excitation current must drop if a capacitive load needs to be supplied. Further literature includes, for example, Boehm [11], Vogel [12], Schroeder [13].

An interesting experiment was undertaken by Köhler [8] to replace the gearbox of a Francis turbine with a very slow-running, permanent magnet excited synchronous generator (Steiner Säge in Hüfingen).

References

1. Beitz W, Küttner KH (1983) Taschenbuch für den Maschinenbau. Springer, Berlin
2. www.bega-wasserkraft.de (2014)
3. www.hydrowatt.de (2014)
4. Henne H (1903) Die Wasserräder und Turbinen – ihre Berechnung und Konstruktion. Verlag von Bernh. Friedr. Voigt, Leipzig
5. Müller W (1899) Die eisernen Wasserräder – Berechnung, Konstruktion und Wirkungsgrad. Verlag Moritz Schäfer, Detmold
6. Bach CV (1886) Die Wasserräder. Verlag von Konrad Wittwer, Stuttgart
7. Palffy SO (1994) Wasserkraftanlagen – Klein- und Kleinstkraftwerke, vol 322, 2nd edn. Expert-Verlag, Renningen-Malmsheim
8. Köhler B (1998) Drehzahlvariable, getriebelose Wasserkraftanlage mit permanentmagneterregtem Synchrongenerator und Francisturbine. PhD Thesis, Universität Karlsruhe
9. Friedrich (1998) Tabellenbuch Elektrotechnik Elektronik. Dümmler Verlag Köln
10. Giesecke J, Mosonyi E, Heimerl S (2009) Wasserkraftanlagen – Planung, Bau und Betrieb, 5th edn. Springer, Berlin
11. Böhm W (1996) Elektrische Antriebe. Vogel Buchverlag Würzburg
12. Vogel J (1988) Elektrische Antriebstechnik. Verlag der Technik, Berlin
13. Schröder D (2000) Elektrische Antriebe – Grundlagen. Springer, Berlin
14. Maher P (2000) Pico hydro for village power – a practical design and installation manual for schemes up to 5kW in hilly or mountainous areas
15. Nuernbergk DM, Trostdorf F (2010) Effizienzsteigerung in Energieerzeugungsanlagen durch Stern-Dreieck-Umschaltung. Wasserkraft und Energie 1:53–69

Chapter 8
Design Flow and Examples

8.1 Design Flow

The quantities nominal flow rate Q and head H are to be determined before the calculation. In the best case, the data for Q and H are available as duration characteristics. The determination of these quantities already takes into account the structural conditions of the hydropower site. The calculation and dimensioning of the millrace, the rake and the inlet and outlet of the channel are already completed then.

Now only the range immediately before and shortly after the wheel is considered. The design flow rate Q is used to dimension the waterwheel. The maximum flow rate Q_{max} to be taken up by the overshoot wheel is higher than the design flow rate.

The calculation process divided into sections A (waterwheel dimensions), B (waterwheel efficiency calculation) and C (dimensioning of the drive system) can then be carried out in the following way:

A1. **Determination of the gross head**

The gross head difference is:

$$H_{gross} = H + \frac{c_1^2 - c_4^2}{2 \cdot g}$$

The gross head may only be used for the efficiency calculation, but not for the wheel design. For the dimensioning of the wheel, the geodetic water level difference H is used. Measurements at different flow rates must be carried out to determine the upstream and downstream water levels accurately. Most important is the measurement of the lower water level at the design flow rate.

A2. **Selection of the waterwheel and channel type**

The head difference should be in a range of $H = 2.0 \ldots 12$ m and the nominal flow rate between $Q = 0.01 \ldots 0.6$ m^3/s (for the application

D. M. Nuernbergk, *Overshot Waterwheels*, Springer Tracts in Mechanical Engineering, https://doi.org/10.1007/978-3-031-95385-9_8

range see also Fig. 1.3). Depending on the circumstances, a choice must be made between over-, undershot waterwheels, hydropower screws, or turbines. Prerequisites for this choice are considerations such as heritage preservation, construction effort, costs etc. The channel type is also determined by these prerequisites. The typical specific discharge capacity is about $Q = 0.22\,\mathrm{m}^3/(\mathrm{s}\cdot\mathrm{m})$ but one can deviate from this value in special cases.

A3. **Assumption of the free hanging H_T**
The waterwheel should slightly or not submerged in the tailwater. According to Henne [1], the free hanging should be:

$$H_T = -0.05\ldots0.1\,\mathrm{m}$$

(see Eq. (5.8) and text there). The experiments by Meerwarth [2] show that a certain wading of the overshot wheel is permissible (experiments on the model wheel). According to Meerwarth [2], wading up to:

$$H_T = 0.2\cdot R_a$$

is acceptable. The loss due to free hanging over the year is kept small if one assumes that for the nominal flow rate the tips of the wheel cells just touch the downstream water level. From the bed slope and the cross-section of the channel in the downstream water, the characteristic curve of the water depth versus flow rate in the channel is then calculated according to the Manning formula (see Eq. (2.1) and values given in Table 2.1). However, it is more accurate to determine the tailwater level by conducting discharge experiments.

A4. **Assumption of the peripheral speed u_A**
The experiments by Meerwarth [2] and Staus [3] on the experimental waterwheel have shown that the peripheral speed can be varied within wide limits without significant loss of efficiency (see Figs. 8.2, 8.3, 10.8, and 10.9). A high peripheral speed should be chosen to achieve small torques and narrow wheels. According to Henne [1], good efficiencies are obtained at a peripheral speed:

$$u_a = 1.5\ldots2.2\,\mathrm{m/s}$$

The optimal ratio should be $c_{2C,u} = 2\cdot u_s$. For more details, refer to Fig. 10.9) and the accompanying text. For small head differences, smaller circumferential speeds are chosen. At higher heads, the peripheral speeds should be high. This specification is a first estimate and will be corrected later. Choosing a higher peripheral speed saves costs, as the wheel becomes narrower and the maximum occurring torque decreases (costs of the gearbox). However, the important peripheral speed is more easily obtained by using the empirical relationship according to Eq. (8.2):

$$u_a = \begin{cases} 1.23 \cdot \sqrt{R_a} \cdot \frac{\mathrm{m}^{1/2}}{\mathrm{s}} & \text{for} \quad R_a \leq 3.2\ \mathrm{m} \\ 2.2\ \frac{\mathrm{m}}{\mathrm{s}} & \text{if} \quad R_a > 3.2\ \mathrm{m} \end{cases}$$

A5. **Assumption of the filling factor ε**

The filling level of the cells (see Eqs. (4.17) and (4.15) and accompanying text) should be in the range of:

$$\varepsilon = \frac{1}{4} \dots \frac{1}{2}$$

Larger values of ε reduce the wheel width. However, increased exit losses occur if the filling factor is high. At the same time, however, the entry loss decreases. Water heels with sheet metal cells should have a design filling factor of $\varepsilon \leq 0.5$. Wheels with wooden blades should have smaller filling factors. The experiments carried out by Meerwarth resulted in a filling factor of $\varepsilon = 0.554$, at which an overload of the wheel occurred (see Fig. 10.10). Such a high fill factor must not be used for the design.

A6. **Assumption of the pressure height**

The pressure height h_{D2} must be specified for waterwheels with entry via a sluice gate or tongue inlet. The pressure height is in the range:

$$h_{D2} = 0.3\ \mathrm{m} \dots 0.5\ \mathrm{m}$$

Equations (3.2) and (3.1) are used to determine $Q = f(h_{D2})$ and $c_{2A} = f(h_{D2})$). Larger values of pressure head h_{D2} lead to larger losses. As a guide, the ration between entrance and peripheral speed can also be used to calculate the pressure head. The ratio should be in a range of $\frac{u_a}{c_{2A}} = 0.56 \dots 0.82$ for a sluice gate. When entering through a broad-crested weir followed by a shooting channel, the velocity c_{2A} (the water depth h_{2A}) is determined for a broad-crested weir followed by a channel with slope by solving the energy equation (Eq. (3.8)—graphical method) or with the Δx method (see Eqs. (3.5), (3.7), (3.9), and (3.10) and accompanying text). The Manning coefficient (see Table 2.1) of the arrangement should be high, i.e., one should prefer a steel sheet channel instead of a wooden channel.

A7. **Information on the channel**

The entry into the channel should be well-rounded to keep the entry loss small. To calculate the water jet exiting the channel, the distances of the channel δ_x and δ_z from the vertex S need to be defined. The distance δ_x determines the entry point B into the wheel, which should ideally be at the upper vertex S. The distance of end of the channel from the vertex in the x-direction is therefore in the range $\delta_x = -0.1 \dots -0.3$ m (see Figs. 5.2 and 5.3—consider startup behavior!). The distance between the waterwheel and the channel bottom δz should be $\delta z = 0.03 \dots 0.05$ m for wooden waterwheels and $\delta z = 0.02 \dots 0.04$ m for steel constructions (consider ice formation in winter). Furthermore, the thickness of the inlet metal plate $m = 0.004 \ldots 0.008$ m needs to be considered.

A8. **Calculation of the ring width a**

The rim width a is calculated as (see Eq. (4.5)):

$$a = \frac{1}{6} \cdot (H)^{1/3} \ldots \frac{1}{4} \cdot (H)^{1/3}$$

Common values are in a range of $a = 0.25 \ldots 0.4$ m. For small flow rates, the chosen values of a should be small so that the water in the cells comes to rest as far out as possible (increase of torque).

The use of Fig. 4.4 is also helpful, which establishes a relationship between discharge capacity, wheel radius and rim width.

A9. **Calculation of the wheel radii R_a, R_i, R_k**

The outer and inner wheel radius are calculated with the following expression:

$$R_a = \frac{H - \delta_z - m - h_{D,2} - H_T}{2}$$

$$R_i = R_a - a$$

The side rings are designed to be slightly larger than the outer wheel radius $R_{K1} = R_a + 0.03 \ldots 0.05$ m to achieve a splash water protection. Furthermore, the center of the waterwheel shaft is determined. The coordinates of the point are $M(x = 0, z = UWS + H_T + R_a)$, where UWS is the tailwater level above sea level. Two reference systems are then introduced. On the one hand the upper and lower water level and the center of the waterwheel shaft are to be given in the coordinates of the geodetic system, on the other hand it is advantageous to place the center M on the coordinates $M(x = 0, z = 0)$. All further information is then related to the point M. This ensures that M has a reference to the geodetic network and simplifies the construction (reference to the point $M = (0, 0)$), e.g. placement of the shaft, arrangement of the gearbox, the generator, the channel etc.

A10. **Determination of the speed n**

The rotational speed is determined by:

$$n = \frac{u_a \cdot 60}{2 \cdot \pi \cdot R_a}$$

However, it is desirable to round this value up or down to a whole number in order to realize simple integer gear ratios. However, it is easier to add a two- to three-stage gearbox with a belt drive that precisely sets the desired speed. This allows for a subsequent correction of the waterwheel speed.

A11. **Calculation of the wheel width B**

For non-ventilated wheels, the width B of the wheel is:

$$B = \frac{Q_{max}}{\varepsilon \cdot a \cdot u_a}$$

The wheel width should not be less than $B = 0.3$ m (ventilation) and should not exceed $B = 4$ m. Another possibility for determining the wheel width is given by the equation:

$$B = \frac{Q_{max}}{Q_s}$$

The specific discharge capacity is then $Q_s = 0.22\ \mathrm{m}^3/(\mathrm{s} \cdot \mathrm{m})$ with a ring width of $a = 0.28$ m to be set or one uses Fig. 4.4.

A12. **The width b_0 of the inlet channel**

For non-ventilated wheels, the width b_0 of the inlet channel must be less than the wheel width B:

$$b_0 = B - 0.1\ \mathrm{m} \ldots 0.2\ \mathrm{m}$$

(see Eq. (3.17) and accompanying text). This ensures lateral ventilation of the cells at the entrance. The side plates of the inlet channel have to be extended over the entrance area of the jet as a splash guard. The jet width s_0 is determined by the equation:

$$s_0 = \frac{Q_{max}}{b_0 \cdot c_{2A}}$$

The exit velocity at point A depends on the type of the water supply.

A13. **Determination of the absolute path of the water jet**

The absolute path of the jet and the lower and upper jet boundary lines are calculated for the maximum flow. Then, the length of the arc $\widehat{B_1 B_2}$ is determined by Newton's method or graphically (see Fig. 3.16). The coordinates of the absolute path of the water jet are determined with the equations (see also Eq. (3.16) and accompanying text):

$$x(t) = c_{2A} \cdot \cos(\gamma) \cdot t + x_A$$

$$z(t) = -\frac{g}{2} \cdot t^2 + c_{2A} \cdot \sin(\gamma) \cdot t + z_A$$

The starting point for the calculation is the exit point $A(x_A, z_A)$ with the coordinates:

$$x_A = -\delta_x$$

$$z_A = R_a + \delta_z + m \quad \text{lower limit of the beam}$$

$$z_A = R_a + \delta_z + m + \frac{s_0}{2} \quad \text{centerline of the jet and}$$

$$z_A = R_a + \delta_z + m + s_0 \quad \text{upper limit of the jet}$$

A14. **Preliminary division** $t_{z,a}$

According to Bach [4] the cell division at the outer radius should be:

$$t_{z,a} = \frac{4}{3} \dots \frac{3}{2} \cdot b_S$$

The arc length can be determined graphically or using Newton's method (intersection of the absolute path of the jet with the radius R_a). The application of either method is avoided if one uses the simple relationship for the division by Henne [1] (see Eqs. (4.6) and (4.7)):

$$t_{z,a} = 0.75 \cdot a + 0.1\, m$$

which gives reliable results. The determined values of the division $t_{z,a}$ are preliminary.

A15. **Determination of the number of cells** z

The number of cells is obtained by:

$$z = \frac{2 \cdot \pi \cdot R_a}{t_{z,a}}$$

This value is to be rounded to a whole number and is preliminary. With this result, one uses Table 4.1 to determine the final value of the number of cells z, the number of arms, the division angle γ_t and the final value for the division

$$t_z = \frac{2 \cdot \pi \cdot R_a}{z}$$

The value t_z should not be chosen too tightly to ensure sufficient ventilation of the cell.

A16. **Determination of the shape of the cell**

For sheet metal cells, the following rules apply (see Fig. 4.2 and Eqs. (4.10) and (4.11)):

$$\widehat{J_1 E} = 1.25 \dots 1.3 \cdot t_z$$

$$\overline{JF} = 0.33 \dots 0.4 \cdot a$$

$$R_H = 2 \dots 3 \cdot \overline{FE} = 2 \dots 3 \cdot \sqrt{(\widehat{J_1 E})^2 + (a - \overline{JF})^2}$$

The straight piece FJ is to be replaced by a semicircular arc (full jet deflection) to achieve best efficiency. For wooden blades, the following applies:

$$\widehat{J_1E} = 1.25 \cdot t_z$$

$$\overline{JF} = 0.4 \ldots 0.5 \cdot a$$

Bent or rolled sheet metal cells are preferable compared to wooden cells (lower losses and better discharge capacity for sheet metal cells, but more complex manufacturing). At the end of the cell design procedure one draws the cell to determine the exit losses.

A17. **Determination of the centroid radius**

This radius is calculated as follows:

$$R_s = \sqrt{\frac{Q \cdot 60}{2 \cdot \pi \cdot n \cdot B} + (R_a - a)^2}$$

Now it is easy to determine the peripheral speed in the centroid radius:

$$u_s = R_s \cdot \frac{2 \cdot \pi \cdot n}{60}$$

A18. **Sketching of the Wheel**

The waterwheel can now be drawn with all specified parameters, as the main design parameters are determined.

B1. **Determination of the inlet losses**

The inlet losses are to be determined as follows (see Eqs. (5.1) and (5.2), Figs. 5.2 and 5.3 as well as the Figs. 8.2 and 10.6):

$$h_{V,e2} = H_C - h$$

$$H_C = R_a + \delta z + m + h_{D2} - z_C$$

$$h = K \cdot \frac{\left(c_{2C,u} - u_s\right) \cdot u_s}{g}$$

The factor K is for wooden cells $K = 1$ and for well-rounded metal cells $K = 1.8$. The speed $c_{2C,u}$ has to be determined at point C via the path equation for the absolute jet (Newton's method).

B2. **Determination of the outlet losses**

The sum of the outlet losses $h_{V,a}$ (static losses according to Eq. (5.5) and dynamic losses according to Eq. (5.6) or (5.7)) are determined after the section "Determination of Losses—Method 1" for various flow rates and rotational speeds. The losses are obtained by linear interpolation between the loss heights of adjacent speed values for the rotational speed n of the wheel. The speed is now to be adjusted so that the sum of the losses for large flows is at a minimum.

B3. **Calculation of mechanical losses**

If friction losses for a certain speed are known for the bearings of the waterwheel, the friction torque M_R can be determined. From this, the loss power has to be determined for the chosen speed of the waterwheel:

$$P_{V,mech} = M_R \cdot \frac{2 \cdot \pi \cdot n}{60}$$

How mechanical losses are determined and evaluated by measurement can be found in Chap. 9. Sometimes it is simpler to assume a constant loss of 1 ... 3%.

B4. **Determination of the waterwheel efficiency**

The efficiency of the overshot waterwheel is determined as follows (see Eq. (6.3) and the accompanying text):

$$\eta_{WR} = 1 - \frac{h_{V,e} + h_{V,a}}{H} - \frac{M_R \cdot u_A}{\varrho \cdot g \cdot Q \cdot H \cdot R_a}$$

Tasks $B1$ to $B3$ can also be omitted if a constant efficiency is assumed. An assumption of efficiency values about $\eta = 0.8 \ldots 0.85$ is a good approximation for well-rounded metal cells. The values of $\eta = 0.75 \ldots 0.8$ are used for wooden and bent metal cells. This approach avoids elaborate efficiency calculations.

It has to be pointed out that efficiency calculations avoid overloading the generator and deliver information on the optimal rotational speed.

C1. **Determination of maximum power**

One uses, if the efficiency of the gear system is known for the operating speed (see Eq. (6.1) and the accompanying text):

$$P_{th} = \varrho \cdot g \cdot Q_{max} \cdot H$$

$$P_{WW} = \eta_{WW} \cdot P_{th}$$

$$P_{elec} = \eta_{Gear} \cdot \eta_{gen} \cdot P_{WW}$$

If the manufacturer's specification for the gearbox efficiency deviates from the operating speed, the friction torque $M_{F,Gear}$ is calculated from the manufacturer's specifications and from this information the power loss and the efficiency for the operating speed of the system. The generator has to be properly chosen for the calculated electrical power P_{elec}. Overloading of the generator should not be allowed. If no standard value is found for the calculated electrical power, it is better to choose the next larger generator type to avoid overloading. The choice of the generator is made according to the product lists of the respective manufacturer (see e.g. Tables 7.3 and 7.4).

C2. **Determination of the maximum torque**

The maximum torque delivered by the waterwheel is (see Eq. (6.4) and the accompanying text):

$$M = \frac{60 \cdot P_{WW}}{2 \cdot \pi \cdot n}$$

This torque is to be used for the dimensioning of the gearbox and the shaft. It should be noted that the waterwheel stresses the shaft also due to its own weight. The mass and the weight force of waterwheels can be estimated according to Bach [4] as follows:

$$m_{WW} = 540 \ldots 580 \cdot a \cdot \left(2 \cdot B \cdot R_a\right)^{3/2} \quad \text{in } kg$$

$$F = m_{WW} \cdot g$$

The design of shafts with torsional and bending stress can be found in Dubbel [5]. The type of bearings (e.g. flying or two-sided bearings in the gearbox) must be taken into account. In the case of flying bearings, the bending vibrations must be determined, otherwise the service life of the gearbox is limited. Usually a clutch is introduced between shaft of the waterwheel and the shaft of gear box.

C3. **Determination of the gear ratio**

The gear ratio is determined by the generator speed under full load conditions and the optimal speed of the waterwheel (see Eq. (7.1) and the text there):

$$i = n_{WW} : n_{Gen}$$

This ratio is to be realized by the gearbox. The generator speed in delta connection is calculated with Eq. (7.3) as follows:

$$n_{gen} = \frac{n_{syn}}{2} \pm \sqrt{\frac{n_{syn}^2}{4} + \frac{60 \cdot P_{th} \cdot \eta_{WW} \cdot \eta_{Gear} \cdot n_{syn}}{2 \cdot \pi \cdot \frac{2 \cdot M_K}{s_K}}}$$

For this, the tipping torque M_k and the tipping slip s_k must be known (specifications from the manufacturer, see also example 6.2). If these specifications cannot be obtained, the specifications from Tables 7.3 and 7.4 for the speed of the generator in motor operation n_{mot} at rated power are used and $n_{gen} = 2 \cdot n_{syn} - n_{mot}$ (linear characteristic) is calculated. The rotational speed for a certain electrical power then is:

$$n_{gen} = \frac{P_{elec}}{P_{elec,nom}} \cdot (n_{gen} - n_{syn}) - n_{syn}$$

To determine the speed in star connection, Eq. (7.4) is used.

C4. **Determination of the annual work**

If the data for the flow rate and the head difference are known as continuous duration curves, the efficiencies can be calculated as a duration curve. The design flow rate of a hydropower site with an overshot wheel should be high according to Meerwarth [2]. Usually, 75 to 50 days of full flow rate are assumed. The characteristics of the electrical power are determined from the values above. The annual electrical yield results from the integral of the characteristic curve of the electrical power P_{elec} given by:

$$W = \int_{0}^{t=365} P_{elec} \cdot dt$$

The determination of the annual yield is done, for example, via the graphical method, which is explained in more detail in Nuernbergk [6] or in Example 9.1. One can assume 200...220 full load days for the calculation of the annual work as a rough estimate. This results in:

$$W = 200 \cdot 24 \cdot P_{N,elec} \quad \text{in } Wh$$

The annual work is the starting point for calculating the profitability of a hydropower plant. A cost-benefit analysis must be carried out in all cases to avoid bad investments (refer Weiss [7]).

The calculation process described here may be repeated several times to be able to adjust all wheel parameters with each other. However, depending on the particular situation of the site, the presented design flow is not carved in stone and one can certainly deviate from it (e.g. for higher flow rates than $Q = 0.6\ \text{m}^3/\text{s}$). The calculation process is more to be understood as an assistance to find a optimum design.

Significant simplifications result if one simply fixes the efficiency of the waterwheel $\eta_{WW} = 0.75 \ldots 0.85$ and the peripheral speed. It is important to assume smaller efficiencies for small head differences and wheels made with wooden cells than for large heads and rounded metal cells. The peripheral speed has to be reduced for smaller heads.

In the following examples, two waterwheels are calculated. The first example exemplifies the complete design flow. The second example wheel is examined in more detail in Chap. 9 using measurements.

The accuracy of the height and length specifications in the examples is 1 mm. This accuracy will not be achieved in practice (perhaps with the height, but not with the length specifications). The fitting accuracy is in the range of 5 ... 10 mm. This must be taken into account in the design.

Provide shims, spacers, distance pieces to cope with the tolerances of the materials. One has to handle e.g. the distance between the water supplying metal sheet and the upper vertex of the waterwheel because e.g. wooden cells tend to swelling under the influence of water, etc.

Similarly, the flow speeds are given up to three places after the decimal point. In practice, these can only be determined accurately up to one or at most two places after the decimal point. The same applies to the flow rate and thus for the theoretical power as well as the useful power (specifications in 0.1-kW steps, i.e. accurate to 100 W). All subsequent examples have been calculated with a scientific calculator.

8.2 Design of an Overshot Waterwheel

Example 8.1 **An overshot waterwheel has to be designed for the Büschleb commercial mill in Worbis (Thuringia). In the past an overshot waterwheel was operated at the site. However, the old waterwheel could not be used as a template, as there are no remnants of the wheel, no old documents or construction drawings available. Different water supply options via sluices or open channels are to be examined for this site and all main dimensions of the waterwheel are to be determined. The mill in Worbis is located on the river Wipper. A weir feeds the mill race above the mill. The flow in the mill channel is a maximum of $Q = 0.5\,\text{m}^3/\text{s}$.**

The flow rate and the geodetic head difference used for the design are summarized in Table 8.1. The waterwheel should be as narrow as possible for structural reasons. It should remain less than 2.1 m in overall width. The operator wants to manufacture the wheel rims and the cells himself, as he owns a carpentry shop. Therefore, the shape of the cells must be simple, with bent sheet metal cells to be used for reasons of discharge capacity. Two wheels have to be designed. The efficiency, the wheel dimensions, and the torque of the different variants are compared, as these directly determine the costs (waterwheel shaft, gearbox, sluices, and drive of the sluice, rake). First, the waterwheel with water supply via sluice gate is calculated, using the design flow. The gross head is equal to the geodetic difference between the headwater and tailwater levels (***A1***):

$$H_{gross} = H_{geo} = 306.438 - 298.80 = 7.638\,\text{m}$$

since the velocities in the channel are not yet known. The free hanging is set to (***A3***):

$$H_T = -0.02\,\text{m}$$

The peripheral speed of the wheel is initially set at (***A4***):

$$u_a = 2.0\,\text{m/s}$$

Table 8.1 Information on the waterwheel location Worbis

Parameter	Value	Remark
Design flow rate	$Q = 0.5\,\text{m}^3/\text{s}$	
Geodetic head	$H = 7.638\,\text{m}$	With cleared tailrace
Geodetic elevation of the headwater	306.438 m a.s.l.	
Bottom of the channel at the free run	306.138 m a.s.l.	
Geodetic elevation of the tailwater	298.80 m a.s.l.	
Bottom of the channel in the wheel chamber	298.50 m a.s.l.	

It is set so high because there is a high head difference and the wheel width B should remain small despite the large flow. The fill level in the cells should be in the range of (***A5***):

$$\varepsilon = \frac{1}{4} \cdots \frac{1}{2}$$

Larger values of ε reduce the wheel width. Since the wheel is to be equipped with straight metal cell pieces (sheet metal or wood), a fill factor of $\varepsilon = 0.3$ is used. The pressure height is chosen to be relatively large (***A6***):

$$h_{D,2} = 0.5 \text{ m}$$

to obtain a narrow inlet jet and high exit velocities c_{2A} at point A of the sluice gate. For the apex S, let $\delta_x = -0.35$ m and $\delta_z = 0.02$ m (***A7***). According to Fig. 5.2, there are low losses for $h_{D2} = 0.5$ m at $\delta_x = -0.5$ m. To also ensure safe start-up of the waterwheel, however, δx cannot be chosen too large. Therefore, the point of first interaction is shifted compared to the upper apex of the wheel ($\delta x = -0.35$ m). The value $\delta_z = 0.02$ m is chosen with regard to icing of the waterwheel. The thickness of the inlet plate is assumed to be $m = 0.005$ m. For the rim width, we get (***A8***):

$$a = \frac{1}{6} \cdot H^{1/3} \ldots \frac{1}{4} \cdot H^{1/3} = 0.328 \ldots 0.492\,\text{m}$$

$a = 0.4\,\text{m}$ is chosen to keep the wheel narrow. The outer and inner wheel radiuses are calculated with (***A9***):

$$\begin{aligned} R_a &= \frac{H - \delta_z - m - h_{D,2} - H_T}{2} \\ &= \frac{7.638 - 0.02 - 0.005 - 0.5 - 0.02}{2} = 3.546\,\text{m} \approx 3.55\,\text{m} \\ R_i &= R_a - a = 3.55 - 0.4 = 3.15\,\text{m} \end{aligned}$$

The side rims are $R_{K1} = R_a + 0.05\,\text{m} = 3.6$ m. They are thus already slightly submerged in the tailwater. According to Fig. 4.4, a specific discharge capacity without construction parts of $Q_s = 0.43\ \text{m}^3/(\text{s} \cdot \text{m})$ is obtained. The speed is determined by (***A10***):

$$n = \frac{u_{A,ang} \cdot 60}{2 \cdot \pi \cdot R_a} = \frac{2 \cdot 60}{2 \cdot \pi \cdot 3.55} = 5.38\ 1/\text{min}$$

However, a speed of $n = 6\,\text{min}^{-1}$ is set. The peripheral speed $u_a = \frac{2 \cdot \pi \cdot R_a \cdot n}{60} = 2.23$ m/s is then at the upper limit of permissibility (the empirical equation $u_a = 1.23 \cdot \sqrt{R_a} = 2.32$ m/s). At such a peripheral speed, the dynamic exit losses must be checked. For non-ventilated wheels, the width B of the wheel is given by (***A11***):

$$B = \frac{Q_{max}}{\varepsilon \cdot a \cdot u_A} = \frac{0.5}{0.3 \cdot 0.4 \cdot 2.23} = 1.868 \text{ m}$$

From the specific discharge capacity, we get: $B = Q/Q_s = 0.5/0.43 = 1.16$ m. As a size for the waterwheel width, $B = 1.9$ m is chosen. The waterwheel will surely be able to discharge 0.5 m^3/s. The width of the two side rims is then added. The requirement of a total width $b < 2.1$ m is just met. The width b_0 of the inlet channel is (***A12***):

$$b_0 = B - 0.1\,\text{m} \ldots 0.2\,\text{m} = 1.7 \ldots 1.8\ \text{m}, \ \text{chosen}\ b_0 = 1.75\ \text{m}$$

This ensures lateral ventilation of the cells at the inlet, as a width of 0.075 m is available for ventilation. It is worth considering whether the jet should be split in the middle to ensure better ventilation of the cell in the middle of the wheel. In very wide wheels without ventilation, a breakthrough of air through the water jet is observed. The calculated ratios are no longer maintained and pulsation phenomena are observed (sudden release of air from the cells). The effort for an inlet via a sluice gate then becomes high, as the sluice gate panel has to be divided into two parts. The inlet channel should be extended beyond the entry area of the jet as a splash water protection. Furthermore, this measure prevents an undesired widening of the free-falling jet. The support of the channel must be sufficiently dimensioned, as per meter a water mass of $m = \varrho \cdot b_0 \cdot h_{D2} \cdot 1\,\text{m} = 999.4\,\text{kg/m}^3 \cdot 1.75\,\text{m} \cdot 0.5\,\text{m} \cdot 1\,\text{m} = 874.5\,\text{kg}$ has to be supported. The speed at the outlet of the sluice gate opening and the jet width without contraction $\varphi = 0.95$ are:

$$c_{2A} = 0.95 \cdot \sqrt{2 \cdot g \cdot h_{D2}} = 0.95 \cdot \sqrt{2 \cdot 9.81 \cdot 0.5} = 2.975\,\text{m/s} \quad \text{and}$$

$$s_0 = \frac{Q_{max}}{b_0 \cdot c_{2A}} = 0.096\,\text{m}$$

The jet remains relatively narrow. This is achieved by the large pressure head $h_{D2} = 0.5$ m. To keep the contraction of the jet very small ($\varphi = 0.95$), the sluice gate panel should be well rounded in the direction of flow. If there is no rounding of the leading edge, the jet width will increase to $s_0 = 0.136$ m, where a coefficient of $\psi = 0.62$ was assumed. The absolute jet is calculated for the maximum flow (***A13***) and the length of the inlet arc $\widehat{B_1 B_2}$ is determined by the Newton method or by graphical determination (Fig. 3.16 and Eqs. 3.11, 3.12 or 3.12 with $x_A = \delta x = -0.35$ m and $z_A = R_a + \delta z + m + s_0/2 = 3.623$ m). With the outlet velocity $c_{2A} = 2.975$ m/s from the sluice gate, one can determine the arc width $b_S = 0.236$ m with the above mentioned equations and the radius R_a. According to Bach [4], the cell division at the outer radius (***A14***) should be:

$$t_{z,a} = \frac{4}{3} \ldots \frac{3}{2} \cdot b_S = 0.315 \ldots 0.354\,\text{m}$$

Application of the Newton method is avoided if the relation for the cell division by Henne [1] is used.

$$t_{z,a} = 0.75 \cdot a + 0.1\,\text{m} = 0.75 \cdot 0.4 + 0.1\,\text{m} = 0.4\,\text{m}$$

The larger cell division $t_{z,a} = 0.4$ m should be used to ensure good ventilation of the cell. The cell count is obtained by dividing by (***A15***):

$$z = \frac{2 \cdot \pi \cdot R_a}{t_{z,a}} = \frac{2 \cdot \pi \cdot 3.55}{0.4} = 55.76$$

This value is to be rounded to a whole number, i.e. $z = 56$. With this result Table 4.1 is used to proceed. The cell count remains $z = 56$, the number of arms is 8, and the division angle is $\gamma_t = 6.42°$. The final value for the division is therefore:

$$t_z = \frac{2 \cdot \pi \cdot R_a}{z} = \frac{2 \cdot \pi \cdot 3.55}{56} = 0.398\,\text{m}$$

For wooden cells (see Fig. 4.2) the following applies (***A16***):

$$\widehat{J_1 E} = 1.25 \cdot t_z = 1.25 \cdot 0.398 = 0.497\,\text{m}$$

$$\overline{JF} = 0.5 \cdot a = 0.5 \cdot 0.4 = 0.2\,\text{m}$$

This form of the wooden cell should also be given to the bent sheet metal cell. The waterwheel can now be drawn, as all main specifications are determined (***A17***). The waterwheel is shown in Fig. 8.4. The losses and the efficiency are to be determined to dimension the torque and rotational speed of the drive system (tasks ***B1*** to ***B4***). The losses are calculated as follows. First, the centroid radius is determined to:

$$R_s = \sqrt{\frac{Q \cdot 60}{2 \cdot \pi \cdot n \cdot B} + (R_a - a)^2} = \sqrt{\frac{0.5 \cdot 60}{2 \cdot \pi \cdot 6 \cdot 1.9} + (3.55 - 0.4)^2} = 3.216 \text{ m}$$

Now the circumferential speed at the centroid radius can be determined:

$$u_s = R_s \cdot \frac{2 \cdot \pi \cdot n}{60} = 3.216 \cdot \frac{2 \cdot \pi \cdot 6}{60} = 2.021 \text{ m/s}$$

Then the intersection of the centroid circle with the absolute path of the jet is determined (see Example 3.5). The distance of point C from the wheel center M is $z_C = 3.164$ m. After that, the velocities c_{2C} and $c_{2C,u}$ at point C are determined. One obtains $c_{2C} = 4.229$ m/s and $c_{2C,u} = 3.455$ m/s. The entry loss is now to be determined as follows (***B1***):

$$H_C = R_a + \delta z + m + h_{D2} - z_C = 3.55 + 0.02 + 0.005 + 0.5 - 3.164 = 0.911 \text{ m}$$

$$h = K \cdot \frac{\left(c_{2C,u} - u_s\right) \cdot u_s}{g} = 1 \cdot \frac{(3.455 - 2.021) \cdot 2.021}{9.81} = 0.295 \text{ m}$$

$$h_{V,e2} = H_C - h = 0.911 - 0.295 = 0.616 \text{ m}$$

The factor $K = 1$ is for wooden cells (or for a sheet metal cell formed like a wooden cell), as a vertical impact on the bar blade and thus no deflection occurs. Here the entry losses are only determined for a speed of $n = 6 \text{ min}^{-1}$. Figure 8.2 shows the entry losses for various flows 0.1 ... 0.5 m^3/s and speeds $n = 3 \ldots 15 \text{ min}^{-1}$. The use of rounded sheet metal cells, which achieve a full deflection of the beam, would reduce the inlet loss to $h_{V,e2} = 0.379$ m. This corresponds to an efficiency increase of 3%, as a reduction of $h_V = 0.076$ m in loss height means a one percent increase in efficiency. The sum of the outlet losses $h_{V,a}$ (static plus dynamic losses) is now determined according to method 1 for various flow rates and speeds (***B2***). First, the cell is drawn as shown in Fig. 8.1. Then, the cell area occupied by the water (see also Fig. 8.1 hatched area) is calculated using the equation:

$$A_z = \frac{Q \cdot 60}{n \cdot z \cdot B} = \frac{0.5 \cdot 60}{6 \cdot z \cdot 1.9} = 0.047 \text{ m}^2$$

This area is occupied by the water in the cell at a speed of $n = 6 \text{ min}^{-1}$ before the water exits the cell (this corresponds to a volume of the full cell of 0.0893 m^3 = 90 l, i.e. approx. 9 normal water buckets of 10 L each). A_z increases with increasing flow rate Q and decreases with increasing speed n. The cell is drawn. Now, the angle $\varphi_{2e} = 63.8°$ at which the water exit is finished is determined graphically and the beam bundle of straight lines is drawn as in Fig. 8.1. The angle difference between each line is $\Delta\varphi = 3°$. However, the lines are drawn at a distance of $\Delta\varphi/2 = 1.5°$.

Next, the distances of the respective points P_i to point E from Fig. 8.1 are determined. The distances $\widehat{EP_i}$ are listed in column 2 of Table 8.2. For example, the length of the first line is $\widehat{EP_1} = 0.525$ m. Now the areas A_1 to A_{25} can be determined. For the area A_1, for example, the following applies:

$$A_1 = \frac{\pi \cdot (\widehat{EP_1})^2}{360°} \cdot \Delta\varphi = \frac{\pi \cdot (0.525)^2}{360°} \cdot 3° = 0.00722 \text{ m}^2$$

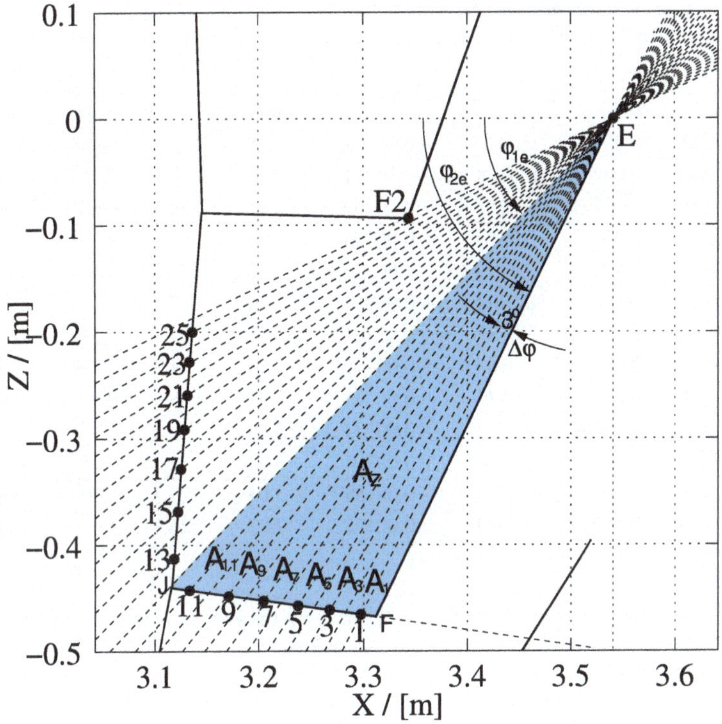

Fig. 8.1 Cell and water filling of the Worbis waterwheel cell

This is $\frac{3}{360}$ of a circle with radius $\widehat{EP_1} = 0.525$ m. Furthermore, the sums of the areas $S_i = \sum_{i=1}^{n} A_i$ with $i = 1, 3, 5, \ldots n$ are also determined. For $i = 9$, the following results:

$$S_9 = A_1 + A_3 + A_5 + A_7 + A_9$$
$$= 0.00722 + 0.00754 + 0.00791 + 0.00835 + 0.00886 = 0.0399 \text{ m}^2$$

Then, the derivatives of the area with respect to the angle are determined. For the area A_1, the following applies:

$$\frac{dA_1}{d\varphi} = -\frac{\pi \cdot (\widehat{EP_1})^2}{360^\circ} = -\frac{\pi \cdot 0.525^2}{360^\circ} = -0.0024 \text{ m}^2/^\circ$$

Now, the speed corresponding to each sum of the areas S_i is determined. This is, for example, the speed n_9, at which the cell volume $B \cdot S_9 = B \cdot (A_1 + A_3 + A_5 + A_7 + A_9)$ would be right before emptying. The formula for $i = 9$ then applies, for example:

$$n_9 = \frac{Q \cdot 60}{S_9 \cdot z \cdot B} = \frac{0.5 \cdot 60}{0.0399 \cdot 56 \cdot 1.9} = 7.066 \text{ min}^{-1}$$

Table 8.2 Calculation results for the losses. The flow corresponds to $Q = 0.5$ m^3/s, i.e. the maximum flow that the waterwheel can handle

i	$\widehat{EP}_i$	A_i	S_i	$\frac{dA_i}{d\varphi}$	n_i	φ_i	$h_{V,a1,i}$	$h_{V,a2,i}$
[1]	[m]	[m^2]	[m^2]	[m^2/°]	[min^{-1}]	[°]	[m]	[m]
1	0.525	0.00722	0.0072	– 0.002410	38.95	62.3	0.405	8.549
3	0.537	0.00754	0.0148	– 0.002514	19.06	59.3	0.451	2.049
5	0.549	0.00791	0.0227	– 0.002635	12.42	56.3	0.501	0.869
7	0.565	0.00835	0.0310	– 0.002784	9.08	53.3	0.555	0.464
9	0.582	0.00886	0.0399	– 0.002955	7.06	50.3	0.613	0.281
11	0.602	0.00950	0.0494	– 0.003167	5.70	47.3	0.675	0.183
13	0.590	0.00912	0.0585	– 0.003039	4.81	44.3	0.736	0.131
15	0.558	0.00814	0.0667	– 0.002713	4.23	41.3	0.793	0.100
17	0.530	0.00735	0.0740	– 0.002451	3.81	38.3	0.848	0.082
19	0.505	0.00667	0.0807	– 0.002223	3.49	35.3	0.901	0.069
21	0.484	0.00613	0.0868	– 0.002045	3.24	32.3	0.954	0.059
23	0.466	0.00570	0.0925	– 0.001899	3.04	29.3	1.007	0.052
25	0.452	0.00534	0.0979	– 0.001782	2.87	26.3	1.056	0.046

The values for the areas A_i, the area sums S_i, the derivatives of the areas with respect to the angle $dA_i/d\varphi$ and the associated speeds n_i are summarized in Table 8.2 in columns 3, 4, 5 and 6. Comparing the area sums S_i with the calculated area $A_z = 0.047$ m^2, it is found that the water surface at a flow rate of $Q = 0.5$ m^3/s must lie between point P_9 and P_{11}. The water discharge begins between point P_9 and point P_{11}.

The corresponding rotational speed lies between $n_{11} = 5.7$ min^{-1} and $n_9 = 7.06$ min^{-1}. The previously set rotational speed was $n = 6$ min^{-1} and lies exactly in the determined range. The cell is only half filled with water as can be seen from Fig. 8.1 (hatched area). The waterwheel is thus not overloaded[1] and the width B of the wheel does not need to be increased. Nevertheless, the use of wood as cell material should be avoided, as the discharge capacity becomes worse and the losses increase unnecessarily. After these preliminary considerations, the exit losses still need to be determined. To calculate the static loss, the angles φ_i for the respective points P_i are needed. The angles[2] are determined with the equation:

$$\varphi_i = \varphi_{2e} - i \cdot \frac{\Delta\varphi}{2} \qquad i = 1, 3, 5, \ldots n$$

$$= 63.8° - 1.5° = 62.3°$$

The angle $\varphi_{2e} = 63.8°$ had already been determined ($\frac{\Delta\varphi}{2} = 1.5°$). This is how the seventh column in Table 8.2 was calculated. The beginning of the water discharge lies between the points P_9 and P_{11}. The starting angle for the water discharge is thus between $\varphi_{1e} \approx 50.3° \ldots 47.3°$. In a range of about $\varphi_{2a} - \varphi_{1e} \approx 13.5° \ldots 16.5°$ the cell is emptied.

[1] Overloading occurs when the water level comes close to point F2 (see Fig. 8.1).

[2] For the definition of the angles φ_{2e}, φ_{1e} see e.g. Fig. 5.6 or 8.1.

The static loss for the corresponding rotational speed n_9 results for example from:

$$
\begin{aligned}
h_{V,a1,n9} &= \frac{1}{S_9} \cdot \Big(A_1 \cdot R_a \cdot (1 - \sin\varphi_1) + A_3 \cdot R_a \cdot (1 - \sin\varphi_3) + \\
&+ A_5 \cdot R_a \cdot (1 - \sin\varphi_5) + A_7 \cdot R_a \cdot (1 - \sin\varphi_7) + \\
&+ A_9 \cdot R_a \cdot (1 - \sin\varphi_9)\Big) \\
&= \frac{3.55}{0.0399} \cdot \Big(0.00723 \cdot (1 - \sin 62.3°) + \\
&+ 0.00754 \cdot (1 - \sin 59.3°) + \\
&+ 0.00791 \cdot (1 - \sin 56.3°) + \\
&+ 0.00835 \cdot (1 - \sin 53.3°) + \\
&+ 0.00886 \cdot (1 - \sin 50.3°)\Big) = 0.613 \text{ m}
\end{aligned}
$$

The loss for the rotational speed $n = 6\ \text{min}^{-1}$ can now be determined from Table 8.2 by linearly interpolating between the values $h_{V,a1.9} = 0.613$ m and $h_{V,a1.11} = 0.675$ m. With the interpolation formula:

$$
\begin{aligned}
y(x) &= \frac{y_{i+1} - y_i}{x_{i+1} - x_i} \cdot (x - x_i) + y_i \\
0.662 &= \frac{0.675 - 0.613}{5.7 - 7.06} \cdot (6 - 5.70) + 0.675
\end{aligned}
$$

Using the values from Table 8.2, we obtain a static loss at a speed of $n = 6\ \text{min}^{-1}$ of $h_{V,a1} = 0.662$ m. To keep the effort small, the dynamic losses are calculated with the equation:

$$
h_{V,a2} = 0.8 \cdot \frac{\pi^2 \cdot R_a^2 \cdot n^2}{1800 \cdot g} = 0.8 \cdot \frac{\pi^2 \cdot 3.55^2 \cdot 6^2}{1800 \cdot 9.81} = 0.203 \text{ m}
$$

For the operation of the waterwheel at a speed of $n = 6\ \text{min}^{-1}$ and a flow rate of $Q = 0.5\ \text{m}^3/\text{s}$ using wooden blades, the following exit losses are obtained:

$$
\begin{aligned}
h_{V,a1} &= 0.662 \text{ m} \\
h_{V,a2} &= 0.203 \text{ m} \\
h_{V,a3} &= H_T = 0.02 \text{ m} \\
h_{V,a} &= h_{V,a1} + h_{V,a1} + h_{V,a1} = 0.89 \text{ m}
\end{aligned}
$$

The results for the loss heights $h_{V,a1,i}$ and $h_{V,a1,i}$ for various associated speeds are summarized in Table 8.2 in the last two columns for a flow rate of $Q = 0.5\ \text{m}^3/\text{s}$. In Figs. 8.2 and 8.3, the losses and efficiency for various flow rates $Q = 0.1 \ldots 0.5\ \text{m}^3/\text{s}$ are plotted over the speed. The calculation was done using a computer program, analogously to the procedure presented here. Such high efficiencies will not be achieved for small flow rates, as the underwater level drops and the loss due to free hanging (set in the figures with $H_T = -0.02\ \text{m} = const$) increases (see also Chap. 9, Fig. 10.25). For well-rounded metal blades, the efficiency could be increased to 85 ... 87% as entry and exit losses become much smaller (almost halved).

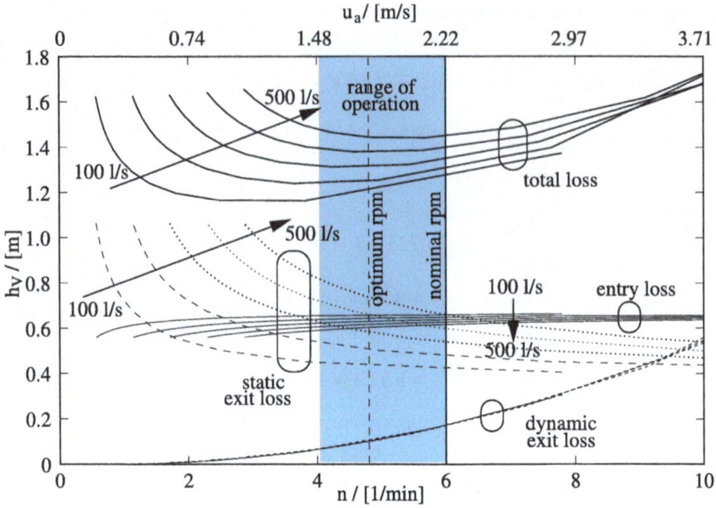

Fig. 8.2 Entry and exit losses at different speeds and flow rates

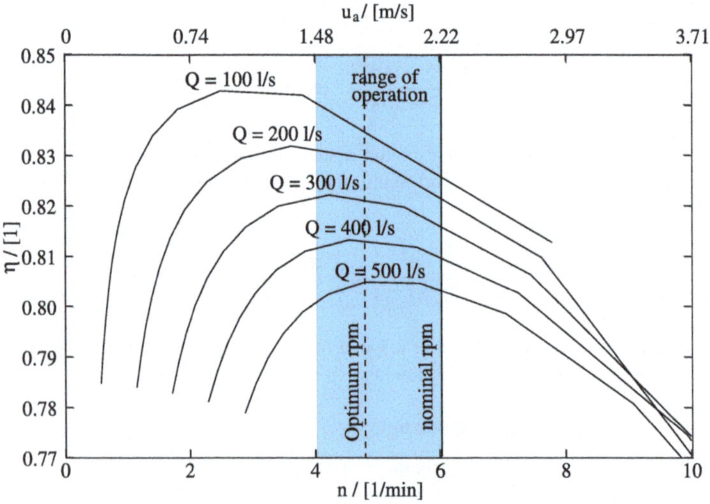

Fig. 8.3 Efficiency for different speeds and flow rates ($H_T = -0.02$ m)

The mechanical loss power in the pendulum roller bearings was neglected, as the details of the bearings were not known at the time of design (***B3***). The waterwheel efficiency at maximum load of $Q_{max} = 0.5$ m^3/s is determined as follows (***B4***):

$$\eta_{WW} = 1 - \frac{h_{V,e} + h_{V,a}}{H} = 1 - \frac{0.617 + 0.89}{7.638} = 0.803$$

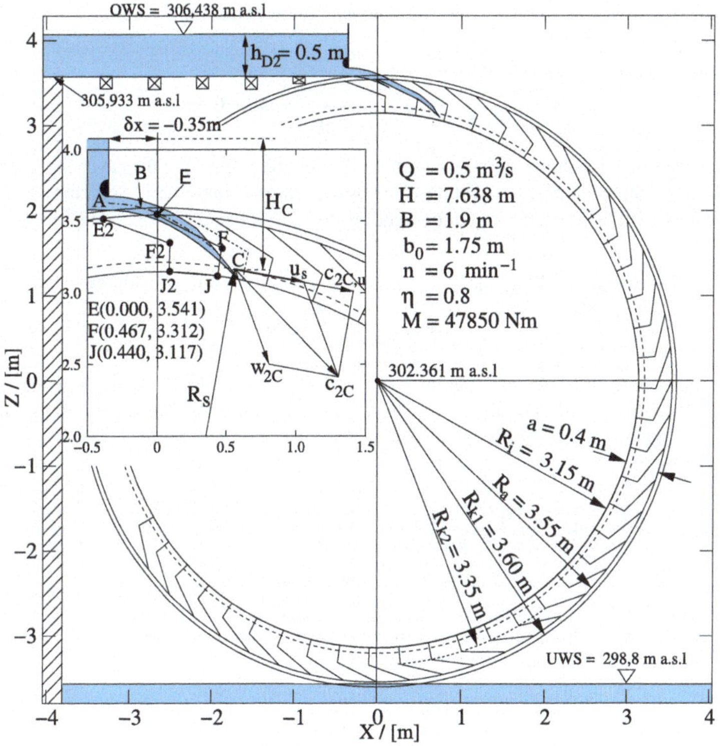

Fig. 8.4 Arrangement of the Worbis waterwheel with feed via a sluice

A reduction in speed would result in an increase in efficiency of less than 1%. Although better efficiencies are achieved for smaller flows, proportionally, larger flows are more frequently processed throughout the year, which also contribute large portions to the total energy. The efficiency increase for small flows does not bring a significant gain in annual energy. A reduction in speed will therefore not be considered, as the torques increase significantly. The mechanical power output from the waterwheel is (***C1***):

$$P_{th} = 9.8 \cdot Q_{max} \cdot H = 9.8 \cdot 0.5 \cdot 7.638 = 37.43 \text{ kW}$$

$$P_{WR} = \eta_{WR} \cdot P_{th} = 0.803 \cdot 37.43 = 30.06 \text{ kW}$$

As gear efficiency, for the three-stage spur gear, $\eta_{Gear} = 0.98^3 = 0.941$ is assumed; the belt drive between the gear block and the generator has an efficiency of $\eta_R = 0.98$ and the high-efficiency generator itself has $\eta_{gen} = 0.935$ (see Table 7.4). This gives us the electrical power and the plant efficiency:

$$P_{elec} = \eta_{gear} \cdot \eta_{gen} \cdot P_{WW} = 0.935 \cdot 0.98^4 \cdot 30.06 = 25.92 \text{ kW} \quad \text{where}$$

$$\eta_{total} = \eta_{WW} \cdot \eta_{gear} \cdot \eta_{gen} = 0.692$$

For the waterwheel, a generator with a rated power of 30 kW will be used, as the next smaller generator of 22 kW would be permanently overloaded. There is also a power reserve if the maximum flow is slightly exceeded. The maximum torque output from the waterwheel is given by (***C2***):

$$M = \frac{60 \cdot P_{WW}}{2 \cdot \pi \cdot n} = \frac{60 \cdot 30060}{2 \cdot \pi \cdot 6} = 47850 \text{ Nm}$$

This torque is used to dimension the waterwheel shaft and is the input torque of the gearbox. For the shaft dimensioning the weight force of the wheel is important. According to Bach [4] the mass and the weight force of overshot waterwheels can be estimated as follows:

$$m_{WW} = 540 \ldots 580 \cdot a \cdot \left(2 \cdot B \cdot R_a\right)^{3/2} = 10700 \ldots 11500 \text{ kg}$$

$$F_G = m_{WW} \cdot g = 105000 \ldots 112800 \text{ N}$$

In this case, an ice layer of 1 … 2 cm thickness must be taken into account when designing the shaft if the wheel should operate in winter time as well. The nominal torque in motor operation of the 30-kW generator is $M_N = 388$ Nm at the nominal speed of $n_N = 740 \text{ min}^{-1}$.

$$M_N = \frac{60 \cdot P_{elec,max}}{2 \cdot \pi \cdot n_N} = \frac{60 \cdot 30000}{2 \cdot \pi \cdot 740} = 387.1 \text{ Nm}$$

see Table 7.4 $\quad M_N = 388$ Nm

The quotient of tipping torque M_k and tipping slip s_k is needed to calculate the generator speed under maximum load. This is obtained as follows:

$$\frac{2 \cdot M_K}{s_K} = \frac{M_N}{1 - \frac{n_N}{n_{sym}}} = \frac{388}{1 - \frac{740}{750}} = 29100 \text{ Nm}$$

where $n_N = 740 \text{ min}^{-1}$ is the nominal speed at nominal torque. The generator rotational speed is calculated using Eq. (7.3) as follows:

$$n = \frac{n_{syn}}{2} + \sqrt{\frac{n_{syn}^2}{4} + \frac{60 \cdot P_{th} \cdot \eta_{WW} \cdot \eta_{Gear} \cdot n_{syn}}{2 \cdot \pi \cdot \frac{2 \cdot M_K}{s_K}}}$$

$$n = \frac{750}{2} + \sqrt{\frac{750^2}{4} + \frac{60 \cdot 37430 \cdot 0.803 \cdot 0.98^4 \cdot 750}{2 \cdot \pi \cdot 29100}} = 759 \text{ min}^{-1}$$

The gear ratio then is (***C3***):

$$i = n_{WR} : n_{Gen,max} = 6 : 759 = 1 : 126.5$$

The translation is to be divided between the two- to three-stage spur gear and the belt drive. The belt drive is used for fine adjustment of the translation. As a rough approximation, 200 full load days are assumed for the calculation of the annual energy. This then gives a value for the annual work (***C4***) of $W = 24 \text{ h} \cdot 200 \cdot 25.9 \text{ kW} = 124320 \text{ kWh}$. It is not recommended to implement the waterwheel with water supply via a sluice gate. This has the following reasons:

1. The supply via a sluice gate does not allow unattended operation, as narrow gate openings tend to clog. A rake must be provided.

2. The weight of the channel to be supported is quite high per meter due to the choice of pressure height $h_{D2} = 0.5$ m. This causes difficulties in the supply up to the upper wheel vertex. The sluice gate plate must be set back (friction losses). The entry losses are therefore high.
3. The sluice gate serves as a control element for the upper water level. The water level control is complex. The water depth must be measured in order to open the gate accordingly. If the gradient in the upper water channel is low, there is significant backwater.
4. In case of emergency closure (electrical energy network not available) the sluice gate must be completely closed to avoid overspeed of the drive system (loss of counter torque). If the gate is already clogged beforehand, safe shutdown is not guaranteed.

A water supply via an open channel has a significantly better performance and is simpler.

8.2.1 Water Supply via an Inclined Open Channel Chute

Example 8.2 **An overshot waterwheel with open channel supply has to be designed for the mill in Worbis (Thuringia). The calculated values should be compared with the results from Example 8.1. All necessary information for the calculation process can be taken from Table 8.1.**

For the supply via an open channel, the design flow does not differ significantly from the already presented process. Differences only arise for the points (**A2**), (**A6**) and (**A7**) of the design flow when the exit velocity from the channel is to be calculated. When supplying a wheel via an open channel, there are two possibilities:

1. Supply via a very long channel section with constant gradient. This case is calculated with the Manning's Eq. (2.1).
2. Supply via a short, steeply inclined channel section (inclined chute). This case is to be calculated like a broad-crested weir. Starting from the critical depth, the sinking line of the water level in the subsequent channel section has to be calculated. The design of the structure can be done with the software WinFlume [8]. At the same time, the flow can be measured relatively accurately ($\pm\, 3 \ldots 6\%$), if a gauge is provided in the upstream water.

Only the second option is further considered. The location details are specified in Table 8.1. The following parameters are also already defined, the free hanging $H_T = -0.02$ m, the peripheral speed $u_a = 2.225$ m/s and the fill factor $\varepsilon = 0.3$, as presented in the previous example (see Table 8.3). The inclined chute is now to be designed. Here, two questions are important for the design:

1. How should the entrance area of the chute be designed to be able to supply the maximum flow $Q = 0.5$ m^3/s to the wheel?
2. What water depth is established after the flow change (from subcritical to supercritical flow) and what speed is reached at the exit point A?

The information from Table 8.1 is important to answer the first question. The water depth at the free run is $h = 0.3$ m. Until the entrance into the chute, the water should be led with a bottom that is 0.1 m deeper. The water depth in front of the chute is thus $h_{Channel} = 0.4$ m. With a flow rate of $Q = 0.5$ m^3/s and a width of the channel of $b_{Channel} = 1.8$ m, the flow speed is about $c_0 = Q/(h \cdot b_{Channel}) \approx 0.7$ m/s. The channel bottom should be filled with medium gravel (see Table 2.2) secured by sole beams. The inlet area is designed like a broad-crested weir. The subsequent channel has an inclination. The definition of all design

Table 8.3 Information on the calculated variants. In the second column of values, only deviating values are entered

Description	Parameter	Sluice gate supply	Open channel water supply	Unit Unit
Maximum flow rate	Q	0.5		[m^3/s]
Geodetic head difference	H	7.638		[m]
Outer wheel radius	R_a	3.54		[m]
Inner radius	R_i	3.14		[m]
Centroid radius	R_s	3.213		[m]
Immersion depth/free hanging	H_T	−0.02		[m]
Rim width	a	0.4		[m]
Filling factor	ε	0.3		[1]
Number of arms	a_z	8		[1]
Wheel width	B	1.9		[m]
Channel width	b_0	1.75	1.8	[m]
Jet width at point A	s_0	0.096		[m]
Impingement arc	b_S	0.236		[m]
Cell count	z	56		[1]
Division	t_z	0.397		[m]
Division angle	γ_t	6.43		[°]
Pressure head	h_{D2}	0.5		[m]
Peripheral speed	u_a	2.225		[m/s]
Circumferential speed in C	u_s	2.019		[m/s]
Rotational speed	n	6		[1/min]
Exit speed at pint A	c_{2A}	2.975	2.894	[m/s]
Efficiency	η	0.802		[1]
Maximum theoretical power	$P_{th,max}$	37,440		[W]
Maximum useful power	$P_{N,max}$	30,020		[W]
Torque at the waterwheel shaft	M	47800		[Nm]
Electric power	P_{elec}	25,890		[W]
Rated power of thr generator	P_N	30		[kW]
Rotational speed of the generator	n_{Gen}	759		[min$^-$1]
Transmission ratio	i	126.5		[1]

parameters used here can be found in Fig. 3.9. The width of the chute should be equal to the channel width in the approach channel. Thus, the width of the chute is $b_0 = 1.8$ m. The theoretical Poleni's coefficient of $C = 1.705\ \mathrm{m}^{1/2}/\mathrm{s}$ is used as a starting value. The overflow height (see also Fig. 3.6) of the broad crested weir then is:

$$h = \left(\frac{Q}{C \cdot b_0}\right)^{2/3} = \left(\frac{0.5}{1.705 \cdot 1.8}\right)^{2/3} = 0.298\ \mathrm{m}$$

The chute is connected to the inflow channel via a rectangular step of 0.1 m. However, the chosen Poleni's coefficient $C = 1.705\ \mathrm{m}^{1/2}/\mathrm{s}$ might be too small and thus more water flows off the free run weir than is actually desired. In order to obtain good flow conditions and a high Poleni's coefficient, the inflow edge must be beveled according to Fig. 3.7 with an angle of $\alpha = 22 \ldots 26°$. One reads from Fig. 3.7 a value $C = 1.98\ \mathrm{m}^{1/2}/\mathrm{s}$ for a ratio of $h/w = 3$. Since $h + w = h_{Channel}$ applies, the equation for the weir height results in:

$$w = h_{Channel} - \left(\frac{Q}{C \cdot b_0}\right)^{2/3} = 0.4 - \left(\frac{0.5}{1.98 \cdot 1.8}\right)^{2/3} = 0.13\ \mathrm{m} \quad \text{and thus}$$

$$h = 0.27\ \mathrm{m}$$

For this ratio $h/w = 2.08$ one obtains the new Poleni's coefficient $C = 1.905$. This value is used for the new calculation of the weir height.

$$w = h_{Channel} - \left(\frac{Q}{C \cdot b_0}\right)^{2/3} = 0.4 - \left(\frac{0.5}{1.905 \cdot 1.8}\right)^{2/3} = 0.123\ \mathrm{m} \quad \text{and}$$

$$h = 0.277\ \mathrm{m}$$

The weir and the overflow height have changed only slightly. The following parameters were chosen $w = 0.125$ m and $h = 0.275$ m. One checks again:

$$w + h = 0.125 + 0.275 = 0.4\ \mathrm{m} \quad \text{and}$$

$$Q = C \cdot b_0 \cdot h^{3/2} = 1.905 \cdot 1.8 \cdot 0.275^{3/2} = 0.494\ \mathrm{m}^3/\mathrm{s}$$

So, the maximum flow rate can be reached by the structure. If the water level now rises above this upper design water level, a free overfall weir limits an further increase of the water depth in the channel. The flow is thus sensibly limited. It should be noted that the water depth in the channel continues to rise, even when the free overfall weir becomes effective (see Fig. 8.5).

In order to discharge a certain flow Q, a corresponding overflow height must also develop at the free flow. Here, a careful hydraulic recalculation is necessary, which depends on the type of overflow edge (broad-crested, circular, sharp-edged etc.) at the free overfall

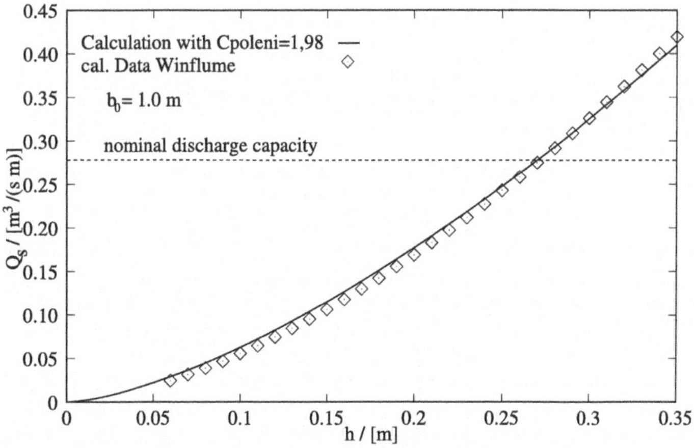

Fig. 8.5 Specific discharge Q_s of the broad-crested weir plotted versus the overflow height h. The data points were calculated with the WinFlume program [8]. The flow rate is obtained with the expression $Q = Q_s \cdot b_0$

Table 8.4 Information on entry via an open chute. See definitions of the information in Fig. 3.9

Description/Parameter	Value	Unit
Water depth in the headwater	0.4	[m]
Velocity of the water in the headwater c_0	0.694	[m/s]
Overflow height h at $Q = 0.5m^3/s$	0.275	[m]
Weir sill height w	0.125	[m]
Inclination Angle of the leading edge α	22°	[1]
Length of the leading edge $\frac{w}{\tan\alpha}$	0.309	[m]
Distance of the gauge level from the leading edge $5 \cdot h$	1.385	[m]
Length of the horizontal weir back $l = 3 \ldots 4 \cdot h$	0.831	[m]
Critical depth on the weir back $h_{crit} = \sqrt[3]{\frac{Q^2}{g \cdot b_0^2}}$	0.199	[m]
Critical speed on the weir back $c_{crit} = \sqrt{g \cdot h_{gr}}$	1.397	[m/s]
Length of the chute L_G	3.5	[m]
Height difference chute δg	0.225	[m]
Gradient chute $I = \frac{\delta g}{L_g}$	0.0643	[1]
Angle deviation from the horizontal $\gamma = -\tan I$	− 3.68°	[1]
Water depth at the exit point A h_{2A}	0.096	[m]
Exit speed at point A $c_{2A} = \frac{Q}{h_{2A} \cdot b_0}$	2.894	[m/s]

weir. An overload of the waterwheel system (mechanical as well as electrical) must be avoided in all cases (Table 8.4).

If, the overflow height at the free run is getting too large due to locally limited conditions, a free run with a sluice gate or a siphon should be considered. For $3 \ldots 4 \cdot h = 0.825 \ldots 1.1$ m the chute channel should run horizontally. Then, the chute should be sloped. A length of 0.9 m is chosen for the horizontal channel piece. The critical depth and the critical speed result in this part of the chute is calculated as:

$$h_{gr} = \sqrt[3]{\frac{Q^2}{g \cdot b_0^2}} = \sqrt[3]{\frac{0.5^2}{9.81 \cdot 1.8^2}} = 0.199 \text{ m}$$

$$c_{gr} = \sqrt{g \cdot h_{gr}} = \sqrt{9.81 \cdot 0.199} = 1.397 \text{ m/s}$$

This jet thickness is still too large and the critical speed too low to supply a waterwheel. The subsequent, inclined channel section therefore has a gradient of 0.225 m over 3.5 m ($I = 0.225/3.5 = 0.0643$) to further accelerate the water and reduce the jet height s_0 further. This is a stationary accelerated flow process (detailed calculation with Δx method, see e.g. Bollrich [9] or Chapter 3). On the horizontal part of the weir back, the emergency closure gate suspended on magnets can be installed, so that in the event of a power failure the waterwheel can be shut down (emergency closure). If a measuring gauge is installed at a distance of about five times the maximum overflow height h in front of the overflow weir, it is easy to determine the flow rate across the structure from Poleni's overflow formula. In the next Fig. 8.5 the results of the WinFlume program are compared with our own calculations (with $C_{Poleni} = 1.905 \text{ m}^{1/2}/\text{s} = const$). The errors calculated by the WinFlume program in the flow rate amount to 2 ... 3%. The agreement would be even better if the overflow coefficient according to Fig. 3.7 had been calculated for the respective ratio of overflow height h to weir height w. This was not pursued further here. A stainless-steel sheet metal should be used for the chute, as one wants to achieve a high Manning-Strickler coefficient $k_{st} = 90\ldots100\,\text{m}^{1/3}/\text{s}$ (see also Table 2.1) to get high exit speeds. With the starting value:

$$h_1 = \left(\frac{Q}{b_0 \cdot k_{st} \cdot I^{1/2}}\right)^{3/5} = \left(\frac{0.5}{1.8 \cdot 90 \cdot 0.0643^{1/2}}\right)^{3/5} = 0.071\ \text{m}$$

Newton's method yields the following for Manning's equation:

$$F(h_i) = Q - k \cdot \frac{h_i^{5/3}}{(2 \cdot h_i + b_0)^{2/3}}$$

$$F'(h_i) = -k \cdot \frac{5/3 \cdot h_i^{2/3} \cdot (2 \cdot h_i + b_0)^{2/3} - 4/3 \cdot h_i^{5/3} \cdot (2 \cdot h_i + b_0)^{-1/3}}{(2 \cdot h_i + b_0)^{4/3}}$$

$$k = b_0^{5/3} \cdot k_{st} \cdot \sqrt{I}$$

$$h_{i+1} = h_i - \frac{F(h_i)}{F'(h_i)}$$

as a solution for the water depth $h = 0.073$ m. The associated velocity is:

$$c_{2A} = \frac{Q}{b_0 \cdot h_0} = 3.805\ \text{m/s}$$

This calculated exit speed is very high. It needs to be checked. For the check, the Δx method can be used, but it is simpler to assume a frictionless case and apply Bernoulli's equation (frictionless fall):

$$h_{Gr} + \frac{c_{Gr}^2}{2 \cdot g} + \delta g = h_{2A} + \frac{c_{2A}^2}{2 \cdot g} \quad \text{and the equation} \quad c_{2A} = \frac{Q}{b_0 \cdot h_{2A}}$$

$$0.523\ \text{m} = 0.199 + \frac{1.397^2}{2 \cdot 9.81} + 0.225 = x + \frac{1}{2 \cdot g} \cdot \left(\frac{0.5}{1.8 \cdot x}\right)^2 = f(x)$$

wheel With the equation above one can calculate the water depth h_{2A} at a certain flow rate Q. The problem is most easily solved graphically. The left side $h_{Gr} + c_{Gr}^2/(2 \cdot g) + \delta g = 0.523$ m is constant and the right side is represented as a function of $h_{2A} = f(x)$.

Both sides are shown together in a graph. As a result in Fig. 8.6, one finds two water depths, which satisfy the equation. Both solutions are valid. The smaller water depth $h_{2A} = 0.096$ m applies to supercritical flow, and the larger water depth $h_{2A} = 0.508$ m applies to subcritical flow. In the channel, however, supercritical flow prevails, because the water depth h_{2A} is smaller than the critical depth. The water depth $h_{2A} = 0.073$ m determined with Manning's equation is too low. One must continue with the water depth determined by the method above, i.g. $h_{2A} = 0.096$ m. This will result in the velocity at the outlet of the channel being significantly lower than calculated with the Manning's equation:

$$c_{2A} = \frac{Q}{b_0 \cdot h_0} = \frac{0.5}{1.8 \cdot 0.095} = 2.894\ \text{m/s}$$

The Manning equation may only be used for very long channels with constant slope, where it can be assumed that a constant, non-accelerated flow process has been established.

The further calculation process for determining the waterwheel data does not deviate from the previously presented example. However, when entering the waterwheel, the non-

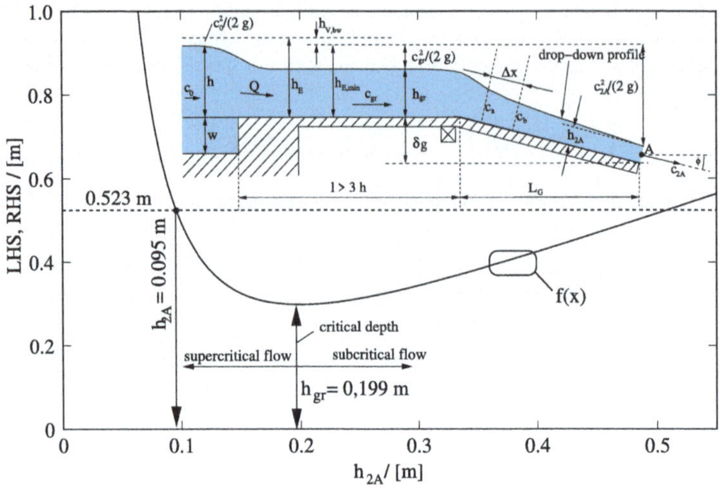

Fig. 8.6 Graphical determination of the water depth at the end of the inclined chute

zero negative starting angle γ must be taken into account in the calculation of the absolute jet (see Fig. 3.15). The angle is $\arctan I = \arctan\,(\delta g/L_G) = \arctan\,(0.225/3.5) = 3.68°$. The last piece of the channel can also be run horizontally, resulting in similar conditions to those when feeding via a sluice gate. This also answers the second question about the water depth and the exit velocity. The waterwheel retains the dimensions calculated in the previous example, as the height of the jet exit when feeding via a channel corresponds to that when feeding via a sluice. Even the exit velocity is approximately the same for a flow rate of $Q = 0.5\ \mathrm{m^3/s}$ at the channel inlet as that obtained via the sluice feed. The waterwheel with water supply over a channel with a chute is shown in Fig. 8.7. The calculated sizes are drawn to scale. The efficiencies for both variants of the feed will differ only slightly, as the velocities at point A do not differ greatly. The differences will be particularly small if a horizontal nozzle follows the section with a slope. Costs are saved with a chute compared to a sluice gate, as a significantly lower water weight has to be supported and no trash rake is needed. The details of the two variants are summarized in Table 8.3. The feed via an open channel is preferable to the feed via a sluice for large head differences as it allows for almost maintenance-free operation of the plant. Sediments and suspended particles can pass through the plant. This also applies to small organisms, which can pass through the waterwheel power plant unharmed. Free passage is ensured in the direction of the tailwater. A rake and a rake cleaning system can be dispensed with if care is taken to keep floating debris which is larger than the opening width between the cells (jet width $s_0 = 0.125$ m) away from the waterwheel. An arrangement similar to the Venturi channel allows precise control of the water flow rate Q supplied to the overshot wheel. In this way, the requirements of the water authorities can be met very precisely and the necessary residual flows can be made available to the natural stream course.

On the other hand, the performance of the waterwheel can be easily checked, whereby with the help of the WinFlume program [8] the errors in the flow measurement can be reduced to below 6%. A regulation of the upper water level by electrical actuators is not necessary. The upper water level adapts to the flow. A disadvantage is the loss of head difference with decreasing water flows. This loss is gladly accepted, as the arrangement remains very simple and low-maintenance.

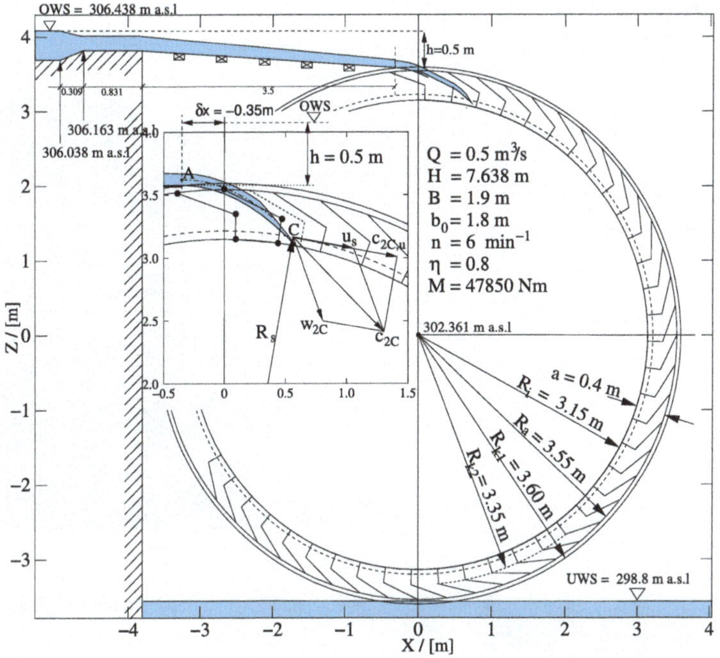

Fig. 8.7 Arrangement of the Worbis waterwheel with open channel feed

Unfortunately, the waterwheel in Worbis could not be constructed in this size, as the tailwater channel was not completely cleared. Therefore, a wheel with a smaller wheel radius and smaller width was used than calculated here in this example. The width and the type of water supply were retained. This wheel and the gear-generator arrangement are shown in Figs. 8.8 and 8.9.

8.3 Overshot Waterwheel in Groß Werzin

Example 8.3 **An overshot waterwheel has to be designed for a hydropower location in Groß Werzin (Brandenburg). The waterwheel is fed from a small lake of the stream Zederbach. The gross head difference at the location is** 3.04 m **and the flow rate is estimated to be** $Q = 300$ l/s. **The flow rate information is not very accurate, and no precise survey of the tailwater level has been conducted. A test with the existing tailwater channel resulted in a strong increase in the underwater level with larger flow rates. This is caused by a narrow pipe which limits the flow in the tailwater area. Furthermore, the slope of the channel bed in the tailwater is very low, so that the water can only flow off at a correspondingly high water depth.**

This flow rate and the information summarized in Table 8.5 serves as a basis for the design. At this point, the simplified design flow is to be used. The design flow is calculated in tabular form (see Table 8.6). Since the flow rate is uncertain, the waterwheel must not be designed with a maximum fill factor of $\varepsilon = 0.5$ (for sheet metal cells).

Fig. 8.8 Overshot waterwheel, Worbis

Fig. 8.9 Gear box, belt drive and generator of the overshot waterwheel, Worbis

Table 8.5 Information on the waterwheel location Groß Werzin

Parameter	Value	Remark
Design flow rate	$Q = 0.3\ \mathrm{m^3/s}$	Value unbound
Geodetic fall height at Q	$H = 3.04\ \mathrm{m}$	

Table 8.6 Simplified design flow at a flow rate of $Q = 0.3\ (0.4)\ m^3/s$

AP	Parameter/Equation	Value calculated	Value selected	Remark
A1.	Gross head difference $H_{Gross} = H_{Geo}$	3.04 m		
A2.	Channel feed	$c_{2A} = 2.137$ m/s		See text
A3.	Free fall H_T		– 0.05 m	UW increase!
A4.	Circumferential speed u_a		1.5 m/s	See text
A5.	Fill factor ε		0.3	
A6.	Pressure height h_{D2}, distance OWS–channel		0.3 m	
A7.	X-distance exit-vertex δx		– 0.15 m	Fig. 5.2
	Z-distance exit-vertex δz		0.02 m	
	Thickness of the channel sheet m		0.005 m	
A8.	Rim width $a = 0.167 \ldots 0.25 \cdot H^{\frac{1}{3}}$	0.24 . . . 0.36 m	0.3 m	
A9.	Outer wheel radius	1.332 m	1.34 m	H_T reduced
	$R_a = \frac{H - \delta z - m - h_{D2} + H_T}{2}$			
	Inner wheel radius			
	$R_i = R_a - a$	1.032 m	1.04 m	
	Radius of side rims			
	$R_K = R_a + 0.05$	1.382 m	1.39 m	Splash water-protection
A10.	Speed $n = \frac{60 \cdot u_a}{2 \cdot \pi \cdot R_a}$, $u_a = 1.5$ m/s	10.69 min^{-1}	10.69 min^{-1}	
A11.	Wheel width $B = \frac{Q}{\varepsilon \cdot a \cdot u_a}$	2.222 m	2 m	
A12.	Channel width $b_0 = B - 0.1 \ldots 0.2$ m	1.8 . . . 1.9 m	1.8 m	
A13.	Absolute path of the jet			See Fig. 8.10
A14.	Pitch $t_{z,a} = 0.75 \cdot a + 0.1$ m	0.325 m		
A15.	Number of cells $z = \frac{2 \cdot \pi \cdot R_a}{t_{z,a}}$	25.9	28	
	$t_z = \frac{2 \cdot \pi \cdot Ra}{z}$	0.303 m		
A16.	Cell shape $\widehat{J_1 E} = 1.25 \cdot t_z$	0.379 m,		
	$\overline{JF} = 0.4 \ldots 0.5 \cdot a$	0.12 . . . 0.15 m	0.12 m	
	$\overline{FE} = \sqrt{(\widehat{J_1 E})^2 + (a - \overline{JF})^2}$	0.42 m	0.42 m	
A17.	Center of gravity radius R_s	1.103 m		
	$R_s = \sqrt{\frac{Q \cdot 60}{2 \cdot \pi \cdot n \cdot B} + (R_a - a)^2}$			
A18.	Drawing of the wheel			See Fig. 8.10
B4.	Efficiency of the waterwheel (WW)		0.8	
C1.	Theoretical power $P_{th} = 9.8 \cdot H \cdot Q$	8.94 (11.92) kW		$Q = 0.3\ m^3/s$
	Mech. power $P_N = P_{th} \cdot 0.8$	7.15 (9.54) kW		Gear. 3-stage
	Elec. power $P_{elec} = P_N \cdot \eta_{gear} \cdot \eta_{gen}$	6.05 (8.07) kW		$\eta_{getr} = 0.94$
	Generator power		11 kW	$\eta_{gen} = 0.9$

(continued)

Table 8.6 (continued)

AP	Parameter/Equation	Value calculated	Value selected	Remark
C2.	Torque of the WW $M = \frac{60 \cdot P_N}{2 \cdot \pi \cdot n}$	6390 (8520) Nm		
C3.	Generator speed			
	$n_{Gen} = n_{syn} - n_{mot}$	785 min^{-1}		From Table 7.4
	Gear ratio $i = n_{WR} : n_{Gen}$	1 : 73.43	1 : 74	
	Gearbox	1 : 25		2-stage
	Belt drive	1 : 2.96		1-stage
C4.	Esti. annual energy $W = 200 \cdot 24 \cdot P_{el}$	29040 kWh		200 days

It is assumed that the inflow could increase to a maximum of $Q_{max} = 400$ l/s. Therefore, a high peripheral speed and a low fill factor $\varepsilon = 0.3$ should be chosen. Furthermore, it should be noted that the generator is chosen large enough that even a flow rate of $Q = 400$ l/s delivers power in its normal working range (under-dimensioning of the generator is not permissible).

It should be noted that the flume supply must be correctly arranged. The calculation of the flume cannot be presented in short form in Table 8.6. For this reason, the flume supply is pre-calculated here. A pressure height of $h_{D2} = 0.3$ m is available to guide the design flow $Q = 0.3$ m^3/s into the wheel.

For a ratio of $h/w = 2$, i.e., the weir sill height is $w = 0.1$ m and the overflow height results in $h = 2 \cdot 0.1$ m $= 0.2$ m, one reads a coefficient of Poleni's formula of $C = 1.9$ m$^{1/2}$/s from Fig. 3.7. It is assumed that the angle of attack of the leading edge is $\alpha = 22 \ldots 26°$ to achieve this high coefficient. The wheel width is preliminarily $B = \frac{Q_{max}}{Q_s} = 0.4/0.2 = 2$ m. The width of the inflow channel then is $b_0 = B - 0.2$ m $= 1.8$ m.

One can now determine the flow rate that can be discharged over the broad-crested weir. With Poleni's formula one obtains:

$$Q = C \cdot b_0 \cdot h^{3/2} = 1.9 \cdot 1.8 \cdot 0.2^{3/2} = 0.306 \text{ m}^3/\text{s}$$

The required water flow rate can be safely supplied to the wheel. If higher flows result, the overflow height in the headwater in the lake upstream will increase. For a flow rate of 400 l/s, the overflow height increases:

$$h = \left(\frac{Q}{C \cdot b_0}\right)^{2/3} = \left(\frac{0.4}{1.9 \cdot 1.8}\right)^{2/3} \approx 0.24 \text{ m}$$

The water level in the lake will therefore rise by the tolerable value of 0.04 m. Now the design of the overflow and the speed at the exit of the arrangement c_{2A} (or the water depth h_{2A}) need to be determined. The critical depth on the weir crest is:

$$h_{gr} = \left(\frac{Q^2}{g \cdot b_0^2}\right)^{1/3} = \left(\frac{0.3^2}{9.81 \cdot 1.8^2}\right)^{1/3} = 0.141 \text{ m}$$

This results in a critical speed of:

$$c_{gr} = \sqrt{g \cdot h_{gr}} = \sqrt{9.81 \cdot 0.141} = 1.176 \text{ m/s}$$

According to the energy theorem (cubic equation):

$$\frac{c_{gr}^2}{2 \cdot g} + h_{gr} + \delta g = \frac{c_{2A}^2}{2 \cdot g} + h_{2A} = \frac{Q^2}{2 \cdot g \cdot h_{2A}^2 \cdot b_0^2} + h_{2A} \quad \text{where} \quad c_{2A} = \frac{Q}{b_0 \cdot h_{2A}}$$

$$0.311 \text{ m} = \frac{0.3^2}{2 \cdot 9.81 \cdot h_{2A}^2 \cdot 1.8^2} + h_{2A}$$

By trying different water depths $h_{2A} < h_{gr} = 0.141$ m, we get:

$$h_{2A} = 0.078 \text{ m} \quad \text{and} \quad c_{2A} = \frac{Q}{b_0 \cdot h_{2A}} = \frac{0.3}{1.8 \cdot 0.078} = 2.137 \text{ m/s}$$

The arrangement of the broad-crested weir is shown in Fig. 8.10. The peripheral speed should be in the range of:

$$u_a = 0.56 \ldots 0.82 \cdot c_{2A} = 1.197 \ldots 1.752 \text{ m/s}$$

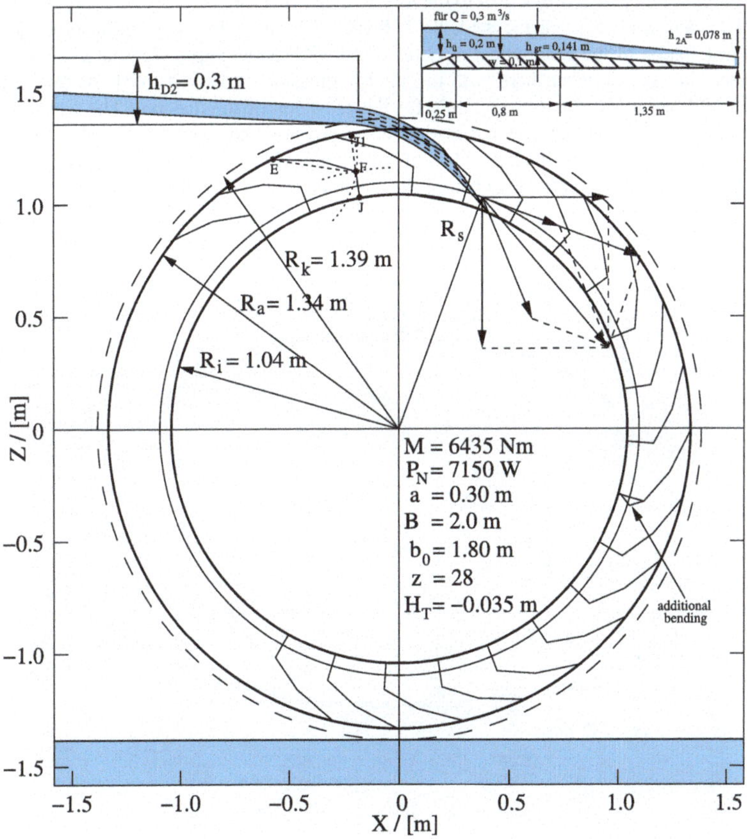

Fig. 8.10 Arrangement of the waterwheel in Groß Werzin with open flume supply

A mean value of $u_a = 1.5$ m/s is chosen (Eq. (8.2) gives $u_a = 1.23 \cdot \sqrt{1.34} = 1.42$ m/s). According to Fig. 4.4, a specific discharge capacity without construction parts of $Q_s = 0.19\ \text{m}^3/(\text{s} \cdot \text{m})$ is obtained, i.e., a wheel width of $B = 0.3/0.19 = 1.57$ m should definitely be exceeded. All further details about the waterwheel are calculated in Table 8.6.

With the information from Table 8.6, the waterwheel can now be drawn. It is divided into two parts due to its large width of $B = 2$ m. The cells are thus no longer than 1 m in the Y direction. This requires a wheel rim in the middle of the wheel. The cells can be offset by 6.44° in the two separated parts to achieve smooth running.

From an assembly point of view, however, the mounting effort would have significantly increased. Therefore, an offset of the cells, although advantageous, was not used. The cell number z was reduced from 30 cells to 28 cells for assembly reasons.

Since the cell shape is made from bent sheet metal, it could deviate from the shape of a wooden cell. The impact blade was bent a second time to improve water entry and exit. In addition, the exit edge is sharp-edged (by bending the sheet metal again at the exit edge) to make it more difficult for the water to exit the cell.

The production remains technologically very simple and yet it achieves higher efficiency compared to a wooden cell. A further improvement would be possible by an additional bending of the impact blade $\overline{FE}$, which would achieve a jet deflection. Due to the low fill factor $\varepsilon = 0.3$ and the relatively high peripheral speed, the cells remain filled with water for a long time on their way down to the lower wheel vertex.

The Figs. 8.11, 8.12, 8.13, 8.14, 8.15, 8.16, 8.17, and 8.18 show the manufacturing and installation of the waterwheel in Groß Werzin. All photographs were taken by Dr. G. Treiber and are displayed here with the kind permission of TWA GmbH. At this point, we would like to express our gratitude for the provision of the material. The waterwheel was completely manufactured and assembled in the workshop of TWA GmbH. Then the prefabricated wheel was loaded and installed on site with relatively low assembly effort. Initial tests of the waterwheel confirmed the calculated ratios (performances in the range of $P_{elec} = 4.5 \ldots 7.0$ kW).

Fig. 8.11 Prefabricated side rim

Fig. 8.12 Fully assembled waterwheel with shaft, hubs, and bearings. With kind permission of TWA GmbH

Fig. 8.13 Fully assembled waterwheel with shaft, hubs, bearings, and claw coupling

At excessively high flow rates Q, the wheel starts to wade, resulting in a drop in performance. This is due to the disturbed outflow conditions in the underwater. This deficiency is to be remedied soon. Since September 1, 2004, the waterwheel in Groß Werzin has been connected to the electrical power grid.

Fig. 8.14 The waterwheel is loaded. With kind permission of TWA GmbH

Fig. 8.15 The waterwheel is installed in the channel

Very similar to this waterwheel design is the overshot waterwheel of the plant in Stedten/Thuringia (see Figs. 10.33 to 10.35). For this wheel, tests have been conducted that confirm a good operational behavior of the values determined here.

Fig. 8.16 The gearbox with two-stage spur gear and single-stage belt drive.With kind permission of TWA GmbH

Fig. 8.17 The waterwheel in operation: view of the inlet

Fig. 8.18 The waterwheel in operation: view of the outlet. With kind permission of TWA GmbH

8.4 Model Scaling Rules for Waterwheels

In recent years, a series of model experiments on waterwheels have been conducted, confirming the good efficiencies of the overshot waterwheel, which have already been reported in older publications [2, 3, 10, 11]. Müller [12] and Becker [13] have conducted extensive investigations of waterwheels with trough flumes. Dubas [14] conducted investigations on the overshot waterwheel, which continue the investigations of Staus [3] and Meerwarth [2]. All these investigations show that overshot waterwheels can be realized with a very good efficiency of over 80% and represent mature constructions.

When reviewing the referenced model experiments, readers often wonder how to apply the impressive results to real-world locations and what scaling rules are required to adapt a model wheel to a full-sized unit for a specific site. The model scaling laws necessary for this task will be derived in this section and their practical application will be explained using the example of the overshot waterwheel [15].

First, the model rules for turbo-machines (see e.g. [9]) are explained and the conversion formulas for the quantities length, force, speed, power, torque and efficiency are determined. Furthermore, a distinction is made between model, large and natural wheel when conversions to a corresponding location are to be made. In the last section, the model laws are applied to the overshot waterwheel.

This enables the derivation of a customized equation for the optimal peripheral speed of an overshot waterwheel, aligning well with model laws and experimental findings. From the optimal peripheral speed, the best rotational speed can also be determined, allowing for the preliminary sizing of drives equipped with an overshot waterwheel. The application of the determined laws is illustrated using an example.

There are various approaches to hydraulic similarities. According to Bollrich [9], one distinguishes:

- the Froude similarity law,
- the Froude similarity law for distorted models,
- the Reynolds similarity law,
- the Weber similarity law,
- the Cauchy-Mach similarity law.

For waterwheels, the Froude similarity law is applied, as one has to describe flow processes with free water surface under atmospheric pressure and discharges over weirs and sluices.

The effects of gravity and inertial forces are particularly significant. The same gravitational acceleration applies and the density of the flowing medium are identical for both the model wheel and the full-scale wheel. Thus, the conversion factor for acceleration is $M_g = 1$ and the conversion factor of the density is given by $M_\varrho = 1$.

Friction processes play a subordinate role for the functionality of the wheels. By choosing a suitable length scale factor M_l, one can scale all sizes accordingly.

8.4.1 *Model Rules for Turbomachines*

Turbomachines are similar to each other if they meet the following three conditions:

1. The machines are geometrically similar. That is the case if all corresponding length dimensions differ by a uniform length scale factor M_l. If we denote the large wheel with G and the model with M, we get:

$$M_l = \frac{L_G}{L_M}$$

2. The machines are kinematically similar, i.e. all motion processes are similar if the speeds of the machines differ by a uniform speed scale factor:

$$M_v = \frac{u_{a,G}}{u_{a,M}} = \frac{c_{2,G}}{c_{2,M}} = \frac{c_{3,G}}{c_{3,M}} = \frac{w_{2,G}}{w_{2,M}} = \frac{w_{3,G}}{w_{3,M}}$$

   This applies to the inlet and outlet speeds c_2, c_3 as well as for the relative speeds w_2, w_3 and the peripheral speeds. Thus, the velocity triangles are also similar if the angles are maintained at $M_\alpha = 1$.

3. The model is dynamically stressed similarly to the large wheel. This requires that all forces and momentum flows differ by a uniform force scale factor:

$$M_f = \frac{F_{m,G}}{F_{m,M}} = \frac{F_{f,G}}{F_{f,M}} = \frac{F_{p,G}}{F_{p,M}} = \frac{\dot{I}_G}{\dot{I}_M}$$

   where F_m denotes a mass force, F_f a friction force, F_p a pressure force and $\dot{I}$ a momentum flow.

The model scale results are expressed as the reciprocal of the scale or conversion factor M_l, written as $1 : M_l$. The proportional reduction of a machine is always possible, so that the first requirement can be well fulfilled. Furthermore, one can adjust the speed and thus the peripheral speed so that the second requirement is well fulfilled. The angular relationships must be maintained in order to achieve similar velocity triangles at the inlet and outlet. The velocities themselves can be suitably adjusted by choosing the volume flows.

Only the dynamic similarity (requirement 3) is not entirely achievable. For the mass and inertia forces (especially weight and centrifugal force), the pressure forces and the momentum flows, one achieves a uniform force scale factor. Only for friction do the force scales differ.

This must be reflected in the conversion of efficiencies so that the efficiencies of the large wheel and the model will differ. However, it is suspected that the differences are small, as friction plays a minor role and most processes in the waterwheel are fully turbulent. Typical friction losses occur, for example, in the overshot wheel as friction of the water in the cells at the inlet and outlet or in the water supply via a chute.

8.4.2 Model Rules for Waterwheels

According to the Froude similarity law, both the large wheel and the model are operated with water and the acceleration due to gravitational acceleration g is the same for both wheels. As previously determined, for the acceleration $M_g = 1$ and for the density $M_\varrho = 1$ apply, while the length follows the length scale factor M_l. All other scale factors for converting waterwheels are given in Table 8.7. The value of the large wheel is to be divided by the scale factor. An arbitrary parameter A of the model wheel can then be determined by the following equation:

$$A_M = \frac{A_G}{M_A}$$

Here, the subscript G stands for large, the subscript M for model wheel, and the size A as well as the associated scale factor M_A can be taken from Table 8.7.

Table 8.7 Conversion with scale factors of the physical values according to Froude's similarity law. The example is valid for a lenth scale factor $M_l = 4$

Parameter	Symbol	Unit	Scale factor	Example
Density	ϱ	[kg/m^3]	$M_\varrho = 1$	$M_\varrho = 1$
Gravity, qcceleration	g, a	[m/s^2]	$M_g = 1$	$M_g = 1$
Length, width, height, pressure height	L, B, H, h_D	[m]	$M_l = M_l$	$M_l = 4$
Vertical and horizontal Surfaces	A	[m^2]	$M_a = M_l^2$	$M_a = 16$
Volume	V	[m^3]	$M_v = M_l^3$	$M_v = 64$
Angle	α	[1]	$M_\alpha = 1$	$M_\alpha = 1$
Number of cells	z	[1]	$M_z = 1$	$M_z = 1$
Time, test time,	t	[s]	$M_t = M_l^{1/2}$	$M_t = 2$
Speed	c, v, u, w	[m/s]	$M_v = M_l^{1/2}$	$M_v = 2$
Rotation speed	n	[1/min]	$M_n = M_l^{-1/2}$	$M_n = 0.5$
Flow rate	Q	[m^3/s]	$M_q = M_l^{5/2}$	$M_q = 32$
Forces, momentum	F	[N]	$M_f = M_l^3$	$M_f = 64$
Power	P	[W]	$M_p = M_l^{7/2}$	$M_p = 128$
Torque	M	[Nm]	$M_M = M_l^4$	$M_M = 256$
Efficiency	η	[1]	$M_\eta = 1$	$M_\eta = 1$

The scale factors for lengths, areas, and volumes are easy to imagine. The scale factors for speed, flow rate, and power are somewhat more complex. These will now be calculated as examples. According to Torricelli's law, the exit speed from an opening is approximately:

$$c = \sqrt{2 \cdot g \cdot h}$$

Thus, the speed can be converted as follows:

$$c = \sqrt{2 \cdot g \cdot h} \sim \sqrt{M_g \cdot M_l} = M_l^{1/2} = M_v$$

The peripheral speed of a waterwheel at point C (see Fig. 8.19) should be approximately half of the inlet speed $u = c/2$. For this reason, the rotational speed converts as follows:

$$n = \frac{u \cdot 60}{2 \cdot \pi \cdot R_a} = \frac{c \cdot 60}{4 \cdot \pi \cdot R_a} \sim \frac{\sqrt{M_l}}{M_l} = M_l^{-1/2} = M_n$$

The flow rate through a cross-sectional area $b \cdot h$ is given by:

$$Q = c \cdot b \cdot h \sim \sqrt{M_l} \cdot M_l^2 = M_l^{5/2} = M_q$$

The theoretical power of a hydroelectric plant is determined by:

$$P = \varrho \cdot g \cdot Q \cdot H \sim M_\varrho \cdot M_g \cdot M_l^{5/2} \cdot M_l = M_l^{7/2} = M_p$$

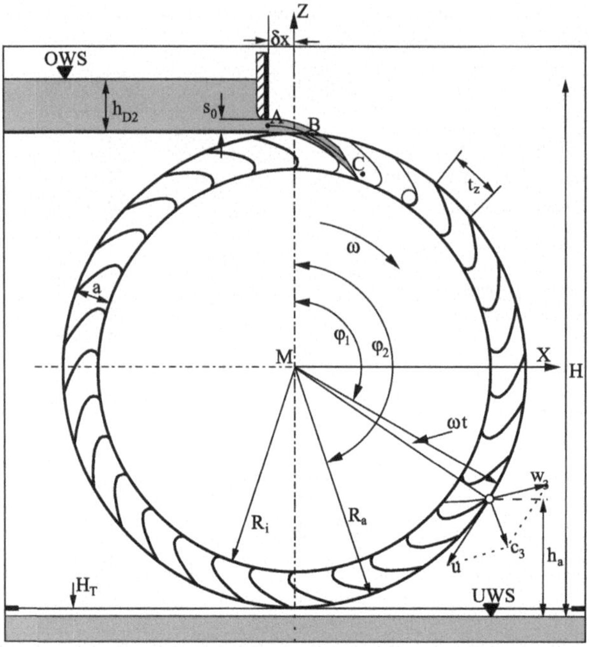

Fig. 8.19 The overshot waterwheel of the test facility used by Staus [3] and Meerwarth [2]

In this way, the conversion factor for power P is obtained. The scale factor of efficiency is:

$$\eta = \frac{P_{out}}{P_{in}} \sim \frac{M_l^{7/2}}{M_l^{7/2}} = M_\eta = 1$$

However, experiments by Meerwarth [2] show that the efficiency of a model waterwheel is slightly greater than that of the large wheel. The differences are still quite small at the optimum point. However, at the boundaries of the operation range (e.g. overspeed, water overload) the model and large wheel differ significantly. The efficiency conversion with $M_\eta = 1$ is therefore only permissible for the optimum point. It is easy to determine the conversion factor of the torque if the conversion of the power is known. This factor is given by:

$$M = \frac{P}{\omega} = \frac{P}{\omega} = \frac{P}{\frac{2\cdot\pi\cdot n}{60}} \sim \frac{M_l^{7/2}}{M_l^{-1/2}} = M_l^4 = M_m$$

It is noticeable that a scaling of the wheel results in a considerable reduction of the torque, which is very desirable for the measurement of the torque on a model wheel. At the end, all important scale factors are known.

8.4.3 Conversion to Natural Waterwheels

By applying the scale factors, one can now:

- Convert large waterwheels into models and thus plan model experiments, efficiency conversion using $M_\eta = 1$.
- Make use of results from model experiments and convert them to large waterwheels.
- Large waterwheels can be transferred to other locations using the rules for the natural wheel.

The scale factor is determined from the geodetic drop height for the model and the large wheel:

$$M_l = \frac{H_G}{H_M}$$

Then, for the conversion of geometric and mechanical quantities, the conversion factors from Table 8.7 are used. If the discharge capacity Q_G of the large wheel derived from the model tests still differs from the desired design flow rate Q_N (where the index N is supposed to stand for nature), the following conversions must be applied:

- For the wheel width of the natural wheel, the design flow rate Q_N is determined as follows:

$$B_N = B_G \cdot \frac{Q_N}{Q_G} \quad \text{with} \quad Q_G = Q_M \cdot M_l^{5/2} \tag{8.1}$$

- for the theoretical and the mechanical useful power of the natural wheel, one obtains:

$$P_{th,N} = \frac{B_N}{B_G} \cdot P_{th,G} \quad \text{and} \quad P_{N,N} = \frac{B_N}{B_G} \cdot P_{N,G}$$

- and for the torque of the natural wheel, the following applies:

$$M_N = \frac{B_N}{B_G} \cdot M_G$$

The use of the equations presented here is demonstrated using examples. It is important that the model wheels must not be chosen too small when planning experimental tests.

8.4.4 Application of the Similarity Relations to Overshot Waterwheels

In Table 8.8, various waterwheels are compiled that only differ by a length scale factor. The large overshot was tested by Staus [3] and Meerwarth [2]. There are two model wheel tests available from literature: there is the model wheel tested in 1935 by Meerwarth [2] and the model tests carried out by Dubas [14] in 2005.

An overshot waterwheel is shown in Fig. 8.19. All dimensions used in Table 8.8 are defined in this figure. From the similarity laws and the previous representation, it can be seen that the circumferential speed at tips of the cells obeys the following similarity:

$$u_{a,opt} \sim \sqrt{M_l}$$

If one chooses the dimension of the wheel radius as the scaling parameter, the optimal circumferential speed can be determined depending on the wheel radius. For the overshot wheel of Staus [3], one gets:

$$u_{a,opt} = 1.65\ \mathrm{m/s} \left(\frac{2 \cdot R_a}{3.6\ \mathrm{m}}\right)^{1/2}$$

From the equation above one obtains:

Table 8.8 Dimensions of the overshot large wheel [3] and the model wheel in a 1:5 scale [2]

Parameter	Size	Large wheel Staus[3]	Model wheel Meerwarth [2]	Model wheel Dubas [14]	Remark
Scale factor	M_l	1	5	4	
Drop height	H	4.0 m	0.8 m	1.0 m	
Flow rate	Q	0.15 $\frac{\mathrm{m}^3}{\mathrm{s}}$	0.0027 $\frac{\mathrm{m}^3}{\mathrm{s}}$	0.0047 $\frac{\mathrm{m}^3}{\mathrm{s}}$	Design flow rate
Outer radius	R_a	1.795 m	0.358 m	0.45 m	
Inner radius	R_i	1.525 m	0.304 m	0.3813 m	
Crown width	a	0.28 m	0.054 m	0.0687 m	
Pitch	t_z	0.376 m	0.075	0.0943	
Number of cells	z	30	30	30	
Wheel width	B	0.8 m	0.1604 m	0.21 m	
Channel width	b_0	0.6 m	0.12 m	–	
Pressure height	h_{D2}	0.4 m	0.08 m	–	
Distance to the sluice	δx	− 0.15 m	–	–	
To the apex	δz	0.017 m	–	–	
Circumferential speed	u_a	1.692 $\frac{\mathrm{m}}{\mathrm{s}}$	0.824 $\frac{\mathrm{m}}{\mathrm{s}}$	0.801 $\frac{\mathrm{m}}{\mathrm{s}}$	At the outer radius
Rotational speed	n	9 min^{-1}	22 min^{-1}	17.0 min^{-1}	Best speed

$$u_{a,opt} = \begin{cases} 1.23 \cdot \sqrt{R_a} \cdot \frac{\text{m}^{1/2}}{s} & \text{for} \quad R_a \leq 3.2 \text{ m} \\ 2.2 \ \frac{\text{m}}{\text{s}} & \text{if} \quad R_a > 3.2 \text{ m} \end{cases} \tag{8.2}$$

The use of the equation for the optimal circumferential speed is only permissible up to a radius of $R_a = 3.2$ m. For larger radii a constant circumferential speed $u_a = 2.2$ m/s is assumed. This was suggested by Bach [4]. With the equation above, the optimal speed $\left(n = \frac{u_a \cdot 60}{2 \cdot \pi \cdot R_a}\right)$ is then also found, namely:

$$n_{opt} = \begin{cases} \frac{11.75 \cdot \frac{\text{m}^{1/2}}{\text{min}}}{\sqrt{R_a}} & \text{for} \quad R_a \leq 3.2 \text{ m} \\ \frac{21.0 \cdot \frac{\text{m}}{\text{min}}}{R_a} & \text{when} \quad R_a > 3.2 \text{ m} \end{cases} \tag{8.3}$$

The optimal ratio between peripheral and inlet velocity was found $u_a/c_{2A} = 0.6$ by Meerwarth [2]. This results in the optimal inlet velocity:

$$c_{2A,opt} = \begin{cases} 2.05 \cdot \sqrt{R_a} \cdot \frac{\text{m}^{1/2}}{s} & \text{for} \quad R_a \leq 3.2 \text{ m} \\ 3.667 \ \frac{\text{m}}{\text{s}} & \text{when} \quad R_a > 3.2 \text{ m} \end{cases} \tag{8.4}$$

From this equation, one can determine the pressure height h_{D2} that is required to supply the water towards the wheel. For a velocity coefficient of $\varphi = 0.95$ (well-rounded exit edge), one obtains:

$$h_{D2} = \begin{cases} \frac{c_2^2}{\varphi^2 \cdot 2 \cdot g} = 0.237 \cdot R_a & \text{for} \quad R_a \leq 3.2 \text{ m} \\ 0.758 \text{ m} \approx 0.7 \text{ m} & \text{when} \quad R_a > 3.2 \text{ m} \end{cases}$$

And as an approximation for the wheel radius ($2 \cdot R_a = H - h_{D2} - \delta_z - H_T - m$, where for simplification, we set $m = 0$ and $H_T = 0$), one then obtains:

$$R_a \approx \frac{H - \delta_z}{2.237}$$

where $\delta_z = 0.02 \ldots 0.05$ m is the distance between the inlet sheet and the wheel.

These equations can now be verified using experiments from [2, 3, 14]. It is also possible to check the theoretical investigations from this book. In Fig. 8.20, the relationships derived above are shown and compared with model, large and natural waterwheels. There is very good agreement between the equations and the experimental data. So, using very simple equations, all essential dimensions of a waterwheel and the optimal operating point can be estimated. This is valuable for the planning of a wheel for a specific hydropower site. Cumbersome calculations of efficiency using the rather unwieldy theoretical model from this book are avoided. In addition, the torque is determined by:

$$M = \frac{P_N}{\omega} = \frac{P_N \cdot 60}{2 \cdot \pi \cdot n}$$

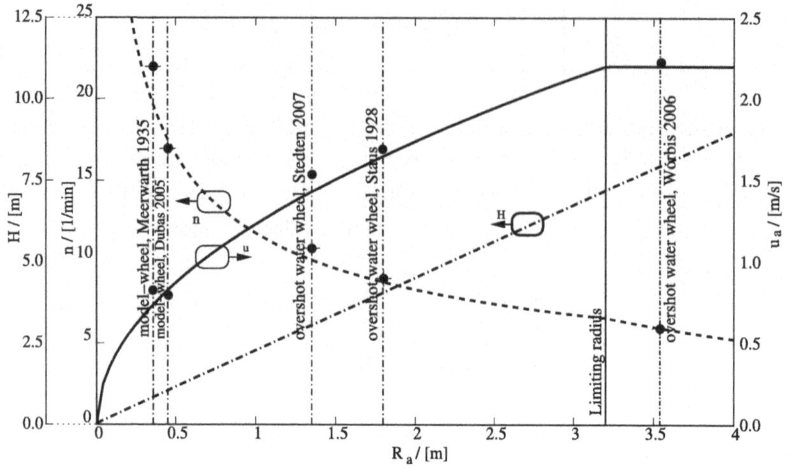

Fig. 8.20 Optimal peripheral speed $u_{a,opt}$, optimal rotational speed n_{opt} and the fall height H plotted over the waterwheel radius R_a. Additionally, the data of the model wheels from Meerwarth [2], Dubas [14], and the large wheel from Staus [3] are shown. Two overshot waterwheels designed in this book (examples Worbis and Werzin) are shown as well. For the last wheels the optimal peripheral speed was determined from theoretical calculations of the waterwheel efficiency

where P_N is the useful power delivered at the waterwheel shaft. As the rotational speed n increases, the torque decreases hyperbolically.

The torque plays a significant role in the costs of the transmission, so that a substantial cost saving can be expected using Eq. 8.3 to determine the optimal speed. The equations above allow for rough planning of a location and quick determination of the project costs of the drive system (wheel, transmission, generator costs). The application of the Eqs. (8.2), (8.3) and (8.4) will now be presented using an example.

Example 8.4 **The geometric dimensions and the speed of an overshot waterwheel with a $H = 5.27$ m and $Q = 0.4$ m are to be estimated. In addition, the large wheel should be converted to the location.**

For the wheel radius, the result is:

$$R_a = \frac{5.27\ \text{m} - 0.05\ \text{m}}{2.237} = 2.33\ \text{m} \approx 2.35\ \text{m}$$

where $\delta_z = 0.05$ m was set. The optimal speed then results in:

$$u_{a,opt} = 1.23 \cdot \sqrt{2.35\ \text{m}} \cdot \frac{\text{m}^{1/2}}{\text{s}} = 1.886\ \frac{\text{m}}{\text{s}}$$

The optimal rotational speed is:

$$n_{opt} = \frac{11.75 \cdot \frac{\text{m}^{1/2}}{\text{min}}}{\sqrt{2.35\ \text{m}}} = 7.66\ \frac{1}{\text{min}}$$

Table 8.9 Conversion of the dimensions of the overshot model wheel [3] to the location. The model scale is 1:1.3175 and the conversion factor of the wheel widths between the large wheel and the natural wheel is 1:1.338. Only the changed sizes of the natural wheel are given

Parameter	Size	Model wheel	Scale-factor	Large wheel	Conversion-factor	Natural Wheel
Head	H	4.0 m	1.3175	5.27	1.0	
Flow rate	Q	0.15 $\frac{m^3}{s}$	1.992	0.299 $\frac{m^3}{s}$	1.338	0.4 $\frac{m^3}{s}$
Outer radius	R_a	1.795 m	1.3175	2.365 m	1.0	
Rim width	a	0.28 m	1.3175	0.369 m	1.0	
Fill factor	ε	0.3	1.0	0.3	1.0	
Pitch	t_z	0.376 m	1.3175	0.495 m	1.0	
Number of cells	z	30	1.0	30	1.0	
Wheel width	B	0.8 m	1.3175	1.054 m	1.338	1.410 m
Channel width	b_0	0.6 m	1.3175	0.790 m	1.338	1.057 m
Pressure height	h_{D2}	0.4 m	1.3175	0.527 m	1.0	
Distance from sluice	δx	− 0.15 m	1.3175	− 0.198 m	1.0	
To apex point	δz	0.017m	1.3175	0.022 m	1.0	
Circumferential speed	u_a	1.692 $\frac{m}{s}$	1.148	1.943 $\frac{m}{s}$	1.0	
Speed	n	9 min^{-1}	0.871	7.84 min^{-1}	1.0	
Theo. power	P_{th}	5884 W	2.625	15445 W	1.338	20665
Efficiency	η	0.84 W	1.0	0.84	1	
Mech. useful power	P_N	4943 W	2.625	12974 W	1.338	17359
Torque	M	5245 Nm	3.013	15802 Nm	1.338	21143

The optimal exit speed at point A is:

$$c_{2A,opt} = 2.05 \cdot \sqrt{2.35\ \text{m}} \cdot \frac{\text{m}^{1/2}}{\text{s}} = 3.143\ \frac{\text{m}}{\text{s}}$$

To achieve this speed with a sluice gate, a pressure height of:

$$h_D = 0.237 \cdot R_a = 0.557\ \text{m}$$

would be necessary. It is then checked whether $2 \cdot R_a + h_D + \delta_z = 2 \cdot 2.35\ \text{m} + 0.557\ \text{m} + 0.05\ \text{m} < H$. If this is not the case, the pressure height is reduced accordingly, in our case by 0.04 m.

A similar result is obtained with the similarity relations from Table 8.7. The conversion of the waterwheel from Staus [3] to the wheel of the site is done using Table 8.9, where first the large wheel is scaled accordingly and then the width of the wheel needs to be adjusted. The model scale for the first conversion results from the ratio of the gradients $M_l = 5.27\ \text{m} / 4\ \text{m}$. Then the discharge capacity or the wheel width is adjusted via the factor $Q_N/Q_G = 1.388$. The agreement between the determined speeds and the speed according to Eqs. (8.2), (8.3), and (8.4) and the information in the table is good.

The inlet width b_0 of the natural wheel should probably be chosen wider, the rim width a should be slightly reduced, and the number of blades z slightly increased. A purely formal application of the similarity relationship is therefore not recommended. A review of the results by the previously specified design flow is certainly useful.

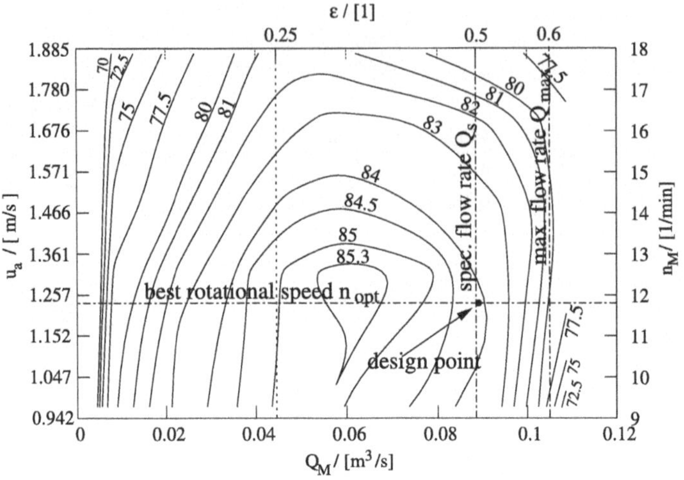

Fig. 8.21 Efficiency contour lines of an overshot waterwheel with the dimensions $R_a = 1$ m, $B = 1$ m and $a = 0.155$ m. The waterwheel is geometrically similar to the wheel of the test facility of Staus [3], but has a comparatively larger width

From measurements using the similarity relations, the efficiency contour lines of a normalized model waterwheel with the dimensions $R_a = 1$ m, $B = 1$ m and $a = 0.155$ m can also be determined. For this purpose, the measurement data from Staus [3] and Meerwarth [2] were combined and presented as a efficiency contour lines diagram. The result is shown in Fig. 8.21.

Additionally, the optimal speed for this wheel of $11.75\ \text{min}^{-1}$, the specific discharge capacity $Q_s = 0.088\ \text{m}^3/(\text{s} \cdot \text{m})$ and the fill factor ε are displayed in the figure. It shows that most of the contour lines are above 80% over a large range of rotational speeds and flow rates. For a given wheel radius, the rotational speed can thus be determined according to Eq. (8.3) or the conversion Table 8.7. The wheel width is determined for a given flow Q_N according to Eq. (8.1) and the efficiency according to Fig. 8.21.

For example, for a location with $H = 3.03$ m, $Q = 0.275\ \text{m}^3/\text{s}$ a wheel radius of $R_a = 1.33$ m is obtained. The conversion factor is then $M_l = 1.33$. Equation (8.3) results in an optimal speed of the natural wheel of $n_N = \frac{11.75}{\sqrt{1.33\ \text{m}}} = 10.2\ \text{min}^{-1}$. The flow of the large wheel results from $Q_G = Q_M \cdot M_q = Q_M \cdot M_l^{5/2} = 0.088 \cdot 1.33^{5/2} = 0.1795\ \text{m}^3/\text{s}$, where the width of the large wheel is $B_G = 1.33$ m and the rim width is $a = 0.206$ m .

The wheel width is thus too narrow to discharge the design flow of 275 l/s. The natural wheel must have a width of $B_N = B_M \cdot M_l \cdot Q_N/(Q_M \cdot M_l^{5/2}) = 1\ \text{m} \cdot 1.33 \cdot 0.275/(0.088 \cdot 1.33^{5/2}) = 2.04$ m according to Eq. (8.1). The efficiency shown in Fig. 8.21 can be used directly for such a wheel, as $M_\eta = 1$ applies.

However, for this case the waterwheel was installed in an old turbine trough with a width of 2 m. Since the bearing had to be taken into account, the wheel could only be built with a maximum width of $B = 1.5$ m. In such cases, the rim width a has to be increased accordingly (refer to Fig. 4.4).

References

1. Henne H (1903) Die Wasserräder und Turbinen – ihre Berechnung und Konstruktion. Verlag von Bernh. Friedr. Voigt, Leipzig (1903)
2. Meerwarth KW (1935) Experimentelle und theoretische Untersuchungen am oberschlächtigen Wasserrad. PhD Thesis, Technische Hochschule Stuttgart
3. Staus A (1928) Wasserradversuche. Wasserkraft und Wasserwirtschaft 18:259–260
4. Bach CV (1886) Die Wasserräder, 1st edn. Verlag von Konrad Wittwer, Stuttgart
5. Beitz W, Küttner KH (1983) Taschenbuch für den Maschinenbau. Springer, Berlin
6. Nuernbergk DM (2005) Wasserräder mit Kropfgerinne – Berechnungsgrundlagen und neue Erkenntnisse, 1st edn. Verlag Moritz Schäfer, Detmold
7. Weiss P (1991) Ein Beitrag zur Planung und Projektierung von Kleinwasserkraftanlagen. PhD Thesis, RWTH-Aachen. Mitteilungen, Institut für Wasserbau und Wasserwirtschaft, RWTH-Aachen
8. Wahl TL, Clemmens AJ, Bos MG, Replogle JA (2013) Winflume, U.S. Department of the Interior, Bureau of Reclamation. www.usbr.gov/pmts/hydraulics_lab/winflume/
9. Bollrich G, Preißler G (1992) Technische Hydromechanik Band 1, 3rd edn. Verlag für Bauwesen, Berlin
10. Müller W (1899) Die eisernen Wasserräder – Berechnung, Konstruktion und Wirkungsgrad. Verlag Moritz Schäfer, Detmold
11. Müller, W (1899) Die eisernen Wasserräder – Berechnung, Konstruktion und Bestimmung ihres Wirkungsgrades, erster Teil – Die Zellenräder, 1st edn. Verlag Veit und Comp.
12. Müller G, Wolters C (2004) The breast shot water wheel: design and model tests. Proc ICE Eng Sustain 157(ES4):203–212
13. Becker T (2004) Physikalische Modellversuche am mittelschlächtigen Wasserrad zur Optimierung der Geometrie und des Wirkungsgrads. Diplomarbeit, Dezember
14. Dubas M (2005) Design and efficiency of overshoot water-wheels. Int J Hydropower Dams 12(3):359–368
15. Nuernbergk DM (2009) Modellgesetze für Wasserräder und ihre Anwendung auf das oberschlächtige Wasserrad. Wasserkraft und Energie 3:38–46

Chapter 9
Measurement Methods

9.1 Introduction

Efficiency studies on waterwheels have a long tradition. For example, one can find measurements of efficiency in the works of Smeaton [1], Zeuner [2], Müller [3], Weidner [4], Staus [5], Meerwarth [6], Neumayer [7], Günther [8], Mueller [9], Dubas [10] and Nuernbergk [11, 12]. The list is far from complete and can certainly be extended (Fig. 9.1).

Extensive early laboratory investigations have been carried out specifically on the overshot waterwheel (Weidner [4], Staus [5], Meerwarth [6] and Dubas [10]). This is probably due to the heavy use and the simple construction of this wheel. Furthermore, relatively small volume flows are needed for such a wheel.

To determine the respective efficiencies, one must independently measure the theoretically available power:

$$P_{th} = \varrho \cdot g \cdot Q \cdot H$$

and the mechanical useful power:

$$P_N = \omega \cdot M \quad \text{with} \quad \omega = \frac{2 \cdot \pi \cdot n}{60}$$

that is delivered by the waterwheel. When feeding the electricity into the grid, the electrical power is also of interest, which is given by:

$$P_{el} = \sqrt{3} \cdot U \cdot I \cdot \cos\phi$$

Therefore the following values must be measured independently to determine the efficiency of a waterwheel:

1. the head difference H,
2. the flow rate Q,

D. M. Nuernbergk, *Overshot Waterwheels*, Springer Tracts in Mechanical Engineering, https://doi.org/10.1007/978-3-031-95385-9_9

(a) (b)

Fig. 9.1 (**a**) Smeaton's [1] measuring apparatus for determining the efficiency of undershot and overshot waterwheels. The time t for lifting a weight of mass m by height h was determined and then the power was calculated with $P = \frac{m \cdot g \cdot h}{t}$. The experiments were published in 1759 and represent the first systematic investigation of the efficiency of overshot and undershot waterwheels. (**b**) An oil painting of John Smeaton by painter Mather Brown, featuring the Eddystone Lighthouse in the background, located in the UK

3. the rotational speed n,
4. the torque M at a point in the drive system and
5. the terminal currents I or voltages U at the generator as well as the power factor ($\cos\varphi$)

A meaningful rounding and error estimation has to be given for each quantity. The measurement methods described in this chapter are necessary to determine the above-mentioned quantities.

9.2 Measurement of the Head Difference

The measurement of the geodetic head is carried out via gauges, which must be calibrated beforehand. As gauges, one uses rods and tracking gauges.

It is useful to install gauges at various points in the channel (e.g. before and after the rake, in the inlet channel, after the water exits the wheel, in the outlet channel).

One should note that, for an accurate calculation of a hydropower plant (e.g., reactivation), the bed slopes in the upper and lower channels, the width of the channels, and the available head differences must also be measured. In other words, the location should be measured as comprehensively as possible.

In doing so, one should look for fixed geodetic points (given in m.a.s.l.—meter above sea level), which ensure a connection to the geodetic land network. All absolute heights are then to be referred to these fixed points. The measurement is best done with a leveling tachymeter and a tape measure; this way, length and height specifications can be correlated.

A very simple method for determining the head difference at complex water structures (water levels before and after buildings) is the use of a hose level.

9.3 Measurement of Flow Rate

For power measurements on waterwheels and other hydraulic machinery, the flows need to be determined very accurately. In general, flow measurements on waterwheels are, in contrast to all other measurements, particularly uncertain [13]. This measurement uncertainty must be taken into account when indicating efficiencies and theoretical performances.

A meaningful rounding and error estimation is to be given after the measurements. The methods generally used for flow measurement are:

1. Water flow measurements with vessels
2. Discharge from openings (Danaids)
3. Weirs measurements
4. Gauge level measurements
5. Measurement with a Venturi channel
6. Measurements with the hydrometric current meter
7. Discharge measurements with the salt dilution method
8. Experiments with depth float and
9. The determination of the precipitation area of the river or stream

These are only the main methods for practitioners, but there are many more. The flows in overshot waterwheels rarely exceed $Q = 1\ \mathrm{m}^3/\mathrm{s}$. Therefore, sufficiently accurate information is obtained with water flow measurements with vessels, weirs, hydrometric current meters and Venturi channels.

For the collection of data of water duration curves, weirs or measurements with a calibrated level at a defined channel section (e.g., Venturi channel) are required. For measurements with a level, it is important that the measurement channel has constant water flow conditions over a longer flow distance. The cross-section perpendicular to the flow direction has to be determined very carefully.

Another method is the determination of the discharge Q from the precipitation area of the river [14]. The salt dilution method is particularly applicable when a channel with predominantly turbulent flow and strong sediment load is present (mountain streams), where the use of alternative methods is limited. It is often useful to combine several of the above methods to determine the flow rate and the nominal flow for a hydropower site.

A very comprehensive presentation of methods for flow measurement in channels with practical examples can be found at the National Bureau of Reclamation [15] or at Herschy [16]. All information in [15] is given in the American unit system and needs to be converted to SI units.

Example 9.1. ***(Determination of a Flow Rate Duration Curve)*** **The flow rate duration curve has to be determined for the river Gera for the period from December 1995 to November 1996. The data is based on the data of the Möbisburg gauge.**

Flow rate duration curves are particularly useful, as they allow precise planning of a hydropower plant. This requires the daily collection of the discharge values of a river or stream over a period of one to several years.

In Table 9.1, you can find the data of the Möbisburg level for the period from December 1995 to November 1996. This data is shown in Fig. 9.2. The area of measured discharges is

Table 9.1 Flow rate values of the river Gera, Möbisburg gauge, December 1995 to November 1996

Day	Dec.	Jan.	Feb.	Mar.	Apr.	May	Jun.	Jul.	Aug.	Sep.	Oct.	Nov.
1	3.07	5.00	2.62	4.20	5.20	6.66	6.24	3.24	3.24	3.42	1.83	2.09
2	2.91	4.60	3.42	3.42	5.00	6.66	5.82	3.42	3.24	3.24	2.09	1.83
3	2.76	4.40	2.62	3.24	4.80	7.10	6.03	3.42	3.07	3.24	1.96	1.71
4	2.76	4.40	2.62	3.07	4.80	6.45	5.61	3.24	3.07	3.07	1.96	1.71
5	2.76	4.00	2.62	3.07	4.80	6.24	5.20	3.42	3.07	3.07	1.96	1.96
6	2.76	3.80	2.62	2.91	4.80	6.24	5.00	3.80	3.24	3.24	1.96	1.83
7	2.62	3.80	2.91	3.07	5.20	6.66	4.80	3.61	3.80	3.42	1.96	1.83
8	2.48	3.80	4.20	2.91	5.40	6.45	4.40	9.90	4.20	2.76	2.22	2.93
9	2.48	3.42	3.80	2.91	6.66	7.35	4.40	12.0	5.20	2.62	2.35	2.63
10	2.48	3.42	3.07	2.91	6.88	7.60	4.20	8.60	5.00	2.62	2.48	2.49
11	2.48	3.24	2.91	2.76	7.10	7.35	4.20	7.10	5.20	2.62	2.35	2.63
12	2.76	3.07	3.42	2.76	8.85	7.10	3.80	6.45	5.82	2.62	2.22	2.36
13	2.76	2.91	4.20	2.76	9.10	7.35	3.42	6.03	7.10	2.76	2.22	2.09
14	2.76	2.91	3.61	2.76	8.10	7.35	3.42	5.82	6.03	2.62	2.09	1.96
15	2.62	2.76	3.07	2.62	7.60	7.10	3.24	5.40	5.61	2.62	2.22	1.83
16	2.62	2.91	3.07	2.62	7.10	7.85	3.07	5.00	5.61	2.35	3.07	3.61
17	2.62	2.91	6.45	2.62	7.10	7.60	3.07	4.60	5.40	2.35	3.80	3.27
18	2.62	2.91	7.60	2.76	7.10	10.5	3.07	4.60	5.40	2.35	3.61	5.82
19	2.62	2.76	6.66	3.24	7.10	11.7	3.07	4.40	4.60	2.35	3.24	6.45
20	2.76	2.62	5.00	4.00	7.60	12.3	3.24	4.20	4.40	2.48	3.24	5.82
21	2.62	2.62	4.00	4.40	8.35	10.8	3.42	4.00	4.60	2.48	5.20	5.82
22	2.76	2.48	3.24	5.00	8.85	9.90	3.24	4.00	4.60	2.35	5.00	5.00
23	6.03	2.48	3.24	6.03	8.10	8.85	3.24	3.80	4.60	2.35	4.60	4.60
24	12.0	2.48	2.91	6.88	11.1	7.85	3.24	4.80	3.24	2.35	4.20	4.60
25	13.8	2.62	3.24	7.35	10.5	7.35	3.24	4.40	3.24	2.35	4.00	3.61
26	9.90	4.00	4.60	7.35	8.35	7.35	3.24	3.80	3.24	2.22	3.61	4.01
27	7.60	4.20	5.20	7.60	7.60	7.35	3.24	3.80	3.24	2.09	3.24	4.01
28	6.66	2.62	4.60	6.88	7.60	7.85	3.07	3.61	4.00	2.09	3.42	2.76
29	5.82	2.48	4.20	6.24	7.85	7.35	3.61	4.00	3.80	2.09	3.42	2.49
30	6.03	2.48	–	5.82	7.25	6.88	3.80	3.61	3.24	1.83	3.42	2.42
31	5.40	2.48	–	5.61	–	6.24	–	3.42	3.07	–	3.24	–

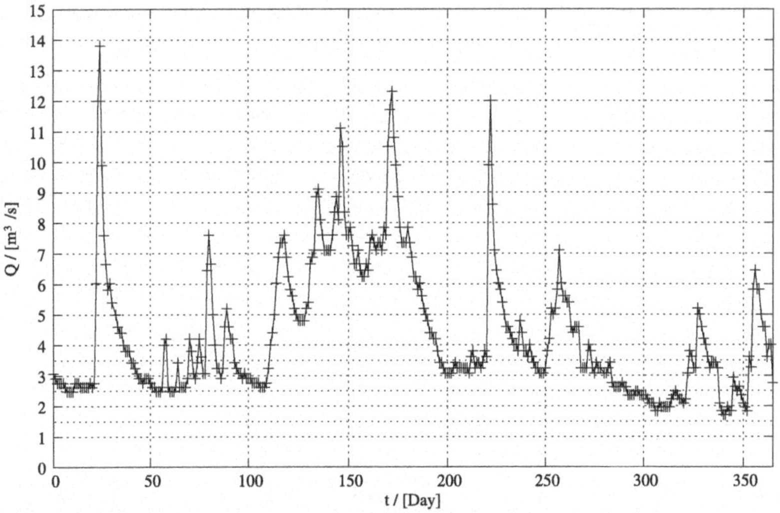

Fig. 9.2 Distribution of flow rate values throughout the year 1996 river Gera

divided into intervals to determine the duration curve, as shown in Fig. 9.2. If a particularly high accumulation of data is found in an interval, it is further subdivided.

Now the intervals are counted as follows: On zero days, the discharge is greater than $Q = 14\ \mathrm{m}^3/\mathrm{s}$. On one day, the discharge is greater than $Q = 13\ \mathrm{m}^3/\mathrm{s}$. On four days, the discharge is greater than $Q = 12\ \mathrm{m}^3/\mathrm{s}$ and so on. This is repeated until one ends the procedure with the value „on 365 days, the discharge is greater than $Q = 1.5\ \mathrm{m}^3/\mathrm{s}$".

In this way, one obtains the values shown in Table 9.2. Column 2 shows the number of days that fall within the respective interval (Column 1). Columns 3 and 4 represent the determined duration curve for the period from December 1995 to November 1996. For comparison, the long-term duration curve for the Möbisburg level is given.

The comparison of the long-term discharge duration line with the duration line of the year shows that even with an observation period of one year, there can be deviations from the long-term average. In 1996, the typical spring flood peak was missing, as there was relatively little snow in the Thuringian Forests. On average, the year was too dry.

The flow duration curve applies to the main river. If the hydropower plant is fed by a diverted flow, the capacity of the millrace determines the usable flow at the hydropower plant. In addition, the residual water flow agreements according to the water law valid for the plant must be taken into account (Fig. 9.3).

The flow of the millrace must be recalculated in such cases. Example 9.1 shows how to plan a hydropower plant with an overshot waterwheel using a flow rate duration curve. Example 2.1 can used for the dimensioning and recalculation of old millraces.

Flow duration curves are recorded in Germany for all larger rivers. These duration lines can be obtained from the water authorities. Many gauge data values (daily flow, average flows) are also available on the internet (e.g., Thuringian State Institute for Environment and Geology).

Table 9.2 Determination of the discharge duration curve of the Gera, Möbisburg level, December 1995 to November 1996. For comparison, the duration line from 1931 to 1996

		Duration line	1996	Duration line	1931–1996
Interval $[m^3/s]$	Days [1]	Days [1]	Q $[m^3/s]$	Days [1]	Q $[m^3/s]$
13.0...14.0	1	1	13.0	0	41.6
12.0...13.0	3	4	12.0	1	34.2
11.0...12.0	2	6	11.0	5	24.1
10.0...11.0	3	9	10.0	10	19.7
9.0...10.0	4	13	9.0	35	12.3
8.0...9.0	8	21	8.0	65	8.85
7.0...8.0	34	55	7.0	95	6.96
6.0...7.0	26	81	6.0	125	5.77
5.0...6.0	33	114	5.0	155	4.86
4.0...5.0	47	161	4.0	185	4.13
3.5...4.0	22	183	3.5	215	3.53
3.0...3.5	69	252	3.0	245	2.96
2.5...3.0	60	312	2.5	275	2.51
2.0...2.5	38	350	2.0	305	2.16
1.5...2.0	15	365	1.5	335	1.86
–	–	–	–	355	1.61
–	–	–	–	360	1.49
–	–	–	–	363	1.40
–	–	–	–	364	1.37

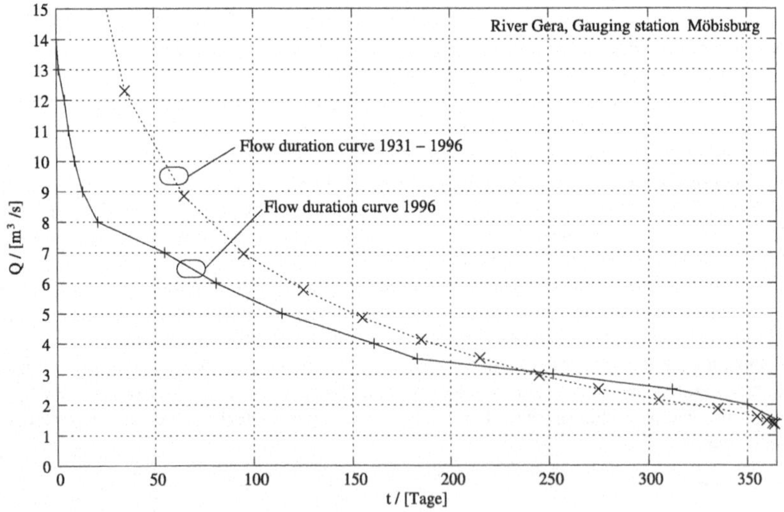

Fig. 9.3 Flow duration curve of the river Gera in 1996 compared to the flow duration curve of the observation period 1931 to 1996

9.3.1 Water Flow Measurements with Water Vessels

Calibration refers to the collection of water in water vessels of known volume content V. The time t is usually measured with a stopwatch. The flow rate is calculated by $Q = V/t$.

This method is only suitable for small flows but is very precise. As with all flow measurements, it is essential to ensure that the measurement is conducted only once a stationary flow state has been reached. Practical hints how to carry out precise measurements can be found in Staus [17] or Herschy [16].

9.3.2 Discharge Measurements with Openings

Water flow measurements can be carried out via corresponding openings. The equations for the exit from gate sluices and bottom and side openings have already been given (compare e.g. Eqs. (3.2) and (3.4)). The outflowing water stream can be inferred by the contraction of the jet in the exit opening of the width s_0.

The problem is that the coefficients to be applied in the equations (contraction of the water jet, velocity reduction) cannot be specified exactly for the respective conditions. For this reason, measurements from openings are relatively inaccurate [18] and calibration is needed (E.g. experiments with Danaids carried out by Staus [17]).

Example 9.2. ***(Flow Measurement with a Sluice Gate)*** **An opening width of $s_0 = 0.08$ m is measured sluice gate with a vertical plate of $b_0 = 2.23$ m width. The water height in front of the sluice gate is determined to be $h_{D2} = 0.47$ m. The flow rate has to be calculated assuming that (a) the inflow edge is well rounded and (b) the inflow edge is sharp-edged.**

The equation for the flow is (3.2):

$$Q = b_0 \cdot \psi \cdot s_0 \cdot c_2 = \frac{\psi}{\sqrt{1 + \frac{\psi \cdot s_0}{h_{D2}}}} \cdot b_0 \cdot s_0 \cdot \sqrt{2 \cdot g \cdot h_{D2}}$$

For the contraction coefficient of a sharp-edged sluice gate, the following applies:

$$\psi(\alpha = 90°) = 1.3 - 0.8 \cdot \sqrt{1 - \left(\frac{\alpha - 205}{220}\right)^2} = 0.618$$

while $\psi \approx 1$ when the sluice gate is very well rounded in the direction of flow. If these values are inserted into the equation for the flow rate, the following values are obtained.

$$\text{(a)} \quad Q = 0.501 \, \frac{m^3}{s} \qquad \psi = 1.0 \quad \text{well rounded}$$

$$\text{(b)} \quad Q = 0.318 \, \frac{m^3}{s} \qquad \psi = 0.618 \quad \text{sharp-edged}$$

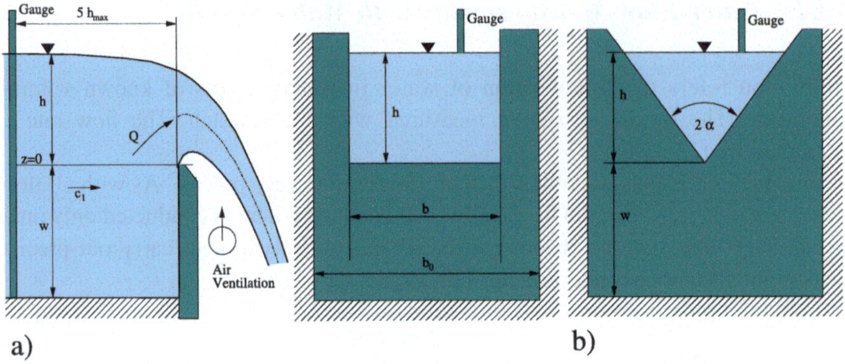

Fig. 9.4 (**a**) Sharp-edged measuring weir with side contraction, (**b**) Triangular measurement weir

The shape of the leading edge of the sluice gate plate has a very large influence on the calculation result. The actual measurement result is probably between these two values. Therefore, only a minimum and maximum value can be given. It should also be noted that the guides of the sluice gate plates often lead to a lateral contraction of the water jet towards the center of the jet. In flow measurements with a submerged sluice gate, the contraction coefficient $\psi = h/s_0$ should also be recorded. Here, s_0 is the opening width of the sluice gate and h is the water depth of the jet emerging from the sluice gate. These values should be measured at a distance of $5 \ldots 8 \cdot s_0$ after the water exit. Error estimation is always needed in flow measurements with sluice gates. (See Herschy [16]).

9.3.3 *Measurement of Discharge with Weirs*

Sharp-edged, horizontal weirs are often used as weirs for discharge measurements, see Bollrich [19]. At a sharp edge, the overfalling jet can detach clearly from the spillway plate. The underside of the jet must be well ventilated in this case.

The spillway coefficient is exactly known under the described circumstances. For sharp-edged rectangular weirs with a horizontal, straight overflow edge, the overflow coefficient μ of the Poleni formula:

$$Q = \mu = \frac{2}{3} \cdot \mu \cdot b \cdot \sqrt{2 \cdot g} \cdot h^{3/2} = 2.953 \cdot \mu \cdot b \cdot h^{3/2}$$

has been determined experimentally with great accuracy.

If the spillway edge is as long as the inlet channel width ($b_0 = b$, see Fig. 9.4), there is no side contraction for spillway. According to experiments by the Swiss Association of Engineers and Architects (SJAV) the coefficient can be calculated by:

$$\mu = 0.615 \cdot \left(1 + \frac{1}{1000 \cdot h + 1.6}\right) \cdot \left[1 + 0.5 \cdot \left(\frac{h}{h + w}\right)^2\right]$$

where the height h is to be inserted in meters. The water flow is then obtained in cubic meters per second. This formula applies under the following conditions:

$$w \geq 0.3\ \text{m} \qquad \frac{h}{w} \leq 1 \qquad 0.025\ \text{m} < h < 0.80\ \text{m}$$

In the case of an existing side contraction ($b < b_0$ as shown in Fig. 9.4a), the spillway coefficient μ of Poleni's formula can be calculated as follows:

$$\mu = \left[0.578 + 0.037\left(\frac{b}{b_0}\right)^2 + \frac{3.615 - 3\cdot\left(\frac{b}{b_0}\right)^2}{1000\cdot h + 1.6}\right]$$
$$\times \left[1 + 0.5\cdot\left(\frac{b}{b_0}\right)^4\cdot\left(\frac{h}{h+w}\right)^2\right]$$

This relationship applies for the conditions:

$$w \geq 0.3\ \text{m} \qquad \frac{b}{w} \leq 1 \qquad 0.025\ \text{m}\cdot\frac{b_0}{b} < h < 0.80\ \text{m}$$

h is to be inserted in meters, and the result is a flow rate in cubic meters per second. Another commonly used measuring spillway is the triangular measuring weir [16, 17, 19] (see Fig. 9.4b). This measuring weir is suitable for smaller discharges, as for example flow rates for overshot waterwheels.

For a central angle of $2\cdot\alpha = 90°$, the discharge is well approximated by the formula (with Q in m^3/s for h in m):

$$Q = 1.352\cdot h^{2.483}$$

It is valid for the range:

$$0.05\ \text{m} \leq h \leq 0.381\ \text{m} \qquad h/w \leq 0.4 \qquad h/b_0 \leq 0.2 \qquad w \geq 0.45\ \text{m}$$

The error of this formula is less than 1% when the above limits are obeyed. It should be noted that the measuring gauge, as shown in Fig. 9.4, should be sufficiently far from the measuring weir, otherwise the decreasing water level line close to the weir will affect the measurement results. The distance to the weir should be greater than five times the maximum spillway height.

The disadvantage of measuring weirs is the need for a specific head difference. On the one hand, this is necessary to achieve a sufficient spillway height. On the other hand, it helps prevent a backwater effect of the downstream water, which could interfere with measurement results.

If the location does not allow for such a head difference, Venturi arrangements or broad-crested weirs (flumes) with a change from subcritical flow to supercritical should be dimensioned using the WinFlume software [20].

*Example 9.3. **(Measuring Weirs)*** **The old water supply channel is to be reactivated for an overshot waterwheel ($H = 6.15$ m). To do the planning the flow rate duration curve has to be determined over a period of one year. The measurement should be carried out with the wooden channel of the old waterwheel supply. The waterwheel was supplied formerly via a sluice gate. The nominal discharge Q is 75 l/s. The width of the wooden channel is $b_0 = 0.5$ m and pressure head is $h_0 = 0.5$ m. How should the measuring channel be designed?**

For the measuring channel, three different measuring weirs are available: (a) a rectangular weir with side contraction, (b) a rectangular weir without side contraction and (c) a triangular weir. The triangular weir cannot be used, as the maximum achievable spillway height is $h = 0.25$ m and thus the maximum measurable discharge $Q = 1.352 \cdot 0.25^{2.483} = 0.043\ \text{m}^3/\text{s}$ is too low.

The triangular weir can only discharge a maximum of 43 liters per second. There is not enough space for the side walls for a rectangular weir with side contraction ($b_1/w > 1$, $0.025 \cdot b_1/b_2 \leq h\ w \geq 0.3$ m), so only the rectangular weir without side contraction can be used. The required conditions $w \geq 0.3$ m and $h/w \leq 1$ are met. To determine the discharge over the measuring weir, the graphical method is chosen (see Fig. 9.5). For this, the weir coefficient for different overflow heights h and $w = 0.3$ m is determined and then the flow is calculated:

$$\mu = 0.615 \cdot \left(1 + \frac{1}{1000 \cdot h + 1.6}\right) \cdot \left[1 + 0.5 \left(\frac{h}{h + w}\right)^2\right]$$

$$Q = \frac{2}{3} \cdot \mu \cdot b_0 \cdot \sqrt{2.0 \cdot g} \cdot h^{1.5}$$

The data is then graphically represented. For the maximum overflow height $h = 0.2$ m the coefficient is obtained:

$$\mu = 0.615 \cdot \left(1 + \frac{1}{1000 \cdot 0.2 + 1.6}\right) \cdot \left[1 + 0.5 \left(\frac{0.2}{0.2 + 0.3}\right)^2\right] = 0.6675$$

$$Q = \frac{2}{3} \cdot 0.6675 \cdot 0.5 \cdot \sqrt{2 \cdot g} \cdot 0.2^{1.5} = 0.088\ \text{m}^3/\text{s} = 88\ \text{l/s}$$

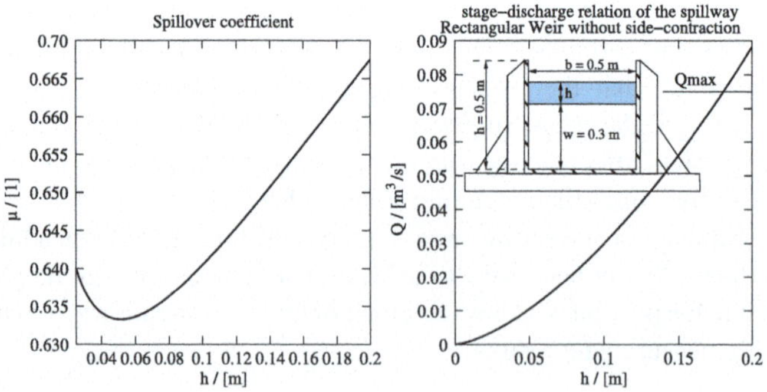

Fig. 9.5 (**a**) Spill coefficient of the sharp-edged measuring weir without side contraction, (**b**) Discharge depending on the overflow height h

From the Fig. 9.5, the weir spillover coefficients and the discharge are read for a given spillway height. In this way the flow rate is easy recorded daily by gauging the spillway height.

9.3.4 Measurement of the Discharge with Venturi Channels or Broad-Crested Weirs (Flumes)

Venturi channels characterized by a narrowed cross-section within open channels. The speed of the water is increased in the narrow cross section. The flow rate Q is determined by measuring the water level difference before and after the Venturi channel.

A Venturi channel is shown in Fig. 9.6. The downstream water does not affect the upstream water level, if a change from sub- to super-critical flow conditions takes place in the Venturi channel. It is then sufficient to measure the level of the upstream water level h_{ow}. The discharge is now calculated with:

$$Q = \mu \cdot C \cdot b_2 \cdot \sqrt{g} \cdot h^{3/2}$$

The coefficient C depends on the shape of the channel and the contraction ratio $m = b_2/b_1$. Information on the coefficients C and μ can be found at Bollrich [19], and detailed information on different Venturi channels (e.g. the Parshall Flume—trapezoidal constriction of the cross-section with or without horizontal sill) can be found at the webpage of the U.S. Department of the Interior, Bureau of Reclamation [15]. Venturi channels have the following advantages:

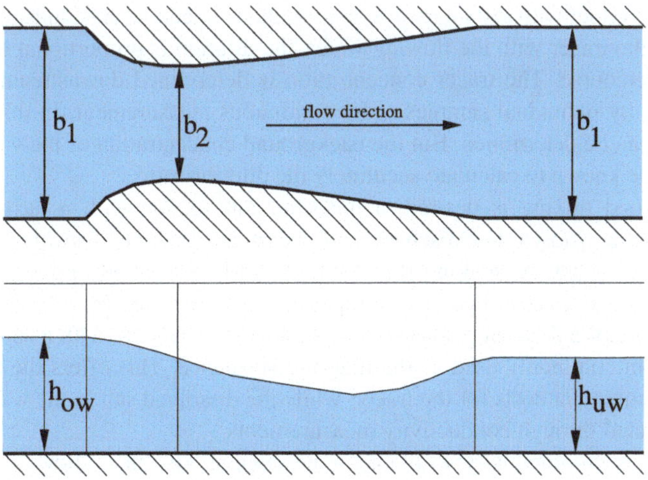

Fig. 9.6 Discharge measurements with the Venturi channel

1. There are no installations in the channel. Thus, there is no obstacle for suspended matter or bulky driftwood.
2. Small to very large discharges can be measured ($Q = 0.001 \ldots 85\ \text{m}^3/\text{s}$).
3. The loss of head is smaller than for example in measurements with overfall weirs.
4. Widths b_1 of a few centimeters up to 15 m can be built.
5. Venturi channels are suitable for automated determination of flow rate duration curves.
6. Portable arrangements can be realized, which can be used in different channels.

The range of application of Venturi channels and broad-crested weirs (flumes) as measurement arrangements is in open irrigation and sewage channels, but can also be used as a feed over an open channel for overshot waterwheels (see Sect. 3.3).

Detailed descriptions of various Venturi channels can be found under the description of Winflume [15]. With the freely available, menu-driven program WinFlume [20], Venturi channels can be easily dimensioned. One advantage of this computer program is that SI- or the American units can be chosen.

9.3.5 Discharge Measurement with the Salt Dilution Gauging Method

In dilution methods, a tracer is introduced into the water flow at a specific point, and its dilution is measured at a cross-section downstream, after passing through the measuring section. This method is mainly used in rivers and streams with high turbulence, very irregular flow cross-section and strong sediment transport. Typical examples are mountain streams and small streams with large gravel.

A solution of the tracer is added to the watercourse in a known concentration. This is done either with instantaneous or continuous injection. Assuming a complete mixing of the tracer with the flowing water, the dilution is proportional to the flow of the watercourse. The tracer concentration is determined downstream from the input point by individual samples or by continuous measurement. In this way, the discharge can be determined. But the background concentration of the watercourse must also be known to calculate accurately the dilution ratio.

Since good mixing is desired, a measurement section with a turbulent flow regime—unlike most other discharge measurement methods—positively impacts measurement accuracy, making this method nearly universally applicable. Electrolytes (salts), dyes or radioactive isotopes are suitable as tracers.

In practice, the dilution method often employs electrolytes, with a sodium chloride solution commonly used as the marking substance. This offers the advantage of low procurement costs for the tracer, while the dissolved salt in the water can be easily detected through conductivity measurements.

There are commercially available experimental kits that allow a discharge measurement with the salt dilution method. Error estimation methods are described in the manual as well (Fig. 9.7).

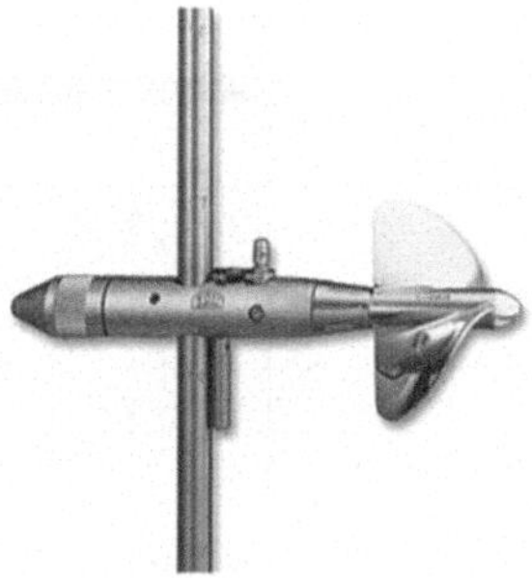

Fig. 9.7 Hydrometric current meter of Ott (With kind permission of OTT HydroMet GmbH, Ludwigstr. 16, D-87437 Kempten)

9.3.6 Measurement of the Discharge with a Hydrometric Current Meter

The determination of the flow rate using the hydrometric current meter is carried out by measuring the flow velocities in a defined cross-sectional area.[1] The water flow condition should be steady for precise measurements.

The cross-sectional area is divided into various sub-areas and the average speed in each area is determined. By summing up the individual flows through the sub-areas, the total flow rate is determined. Changes to the environmental settings by opening sluice gates or bypass channels should be avoided.

An example of the current meter measurement is provided, as described by Neumayer [7]. Here is a verbatim quote translated into English: „The local conditions proved to be ideal for a current meter measurement: rectangular cross-sectional area with vertical masonry walls and an almost flat gravel bed without weed growth. The channel bed was secured at a distance of a few meters by wooden ground peg. An existing bridge could be used as a measuring bridge (Fig. 9.8).

With the current meter measurement, the so-called point measurement method was used; here, the water velocity is determined at individual points of the measuring plumb line. The resulting average velocity then refers to the sub-area (catchment cross-section) of the cross-sectional area associated with the measuring plumb line.

Figure 9.8 also shows the velocity distribution along the four measuring plumb lines. The respective average velocities could be determined by area compensation."

The following values were obtained for the average velocities and the associated sub-areas:

$$c_{mA} = 0.38\ \frac{m}{s} \quad F_A = 0.67 \cdot 1.40 = 0.938\ \text{m}^2$$

$$c_{mB} = 0.41\ \frac{m}{s} \quad F_B = 0.66 \cdot 1.40 = 0.924\ \text{m}^2$$

[1] The hydrometric current meter was invented by Reinhard Woltman in 1790 [21].

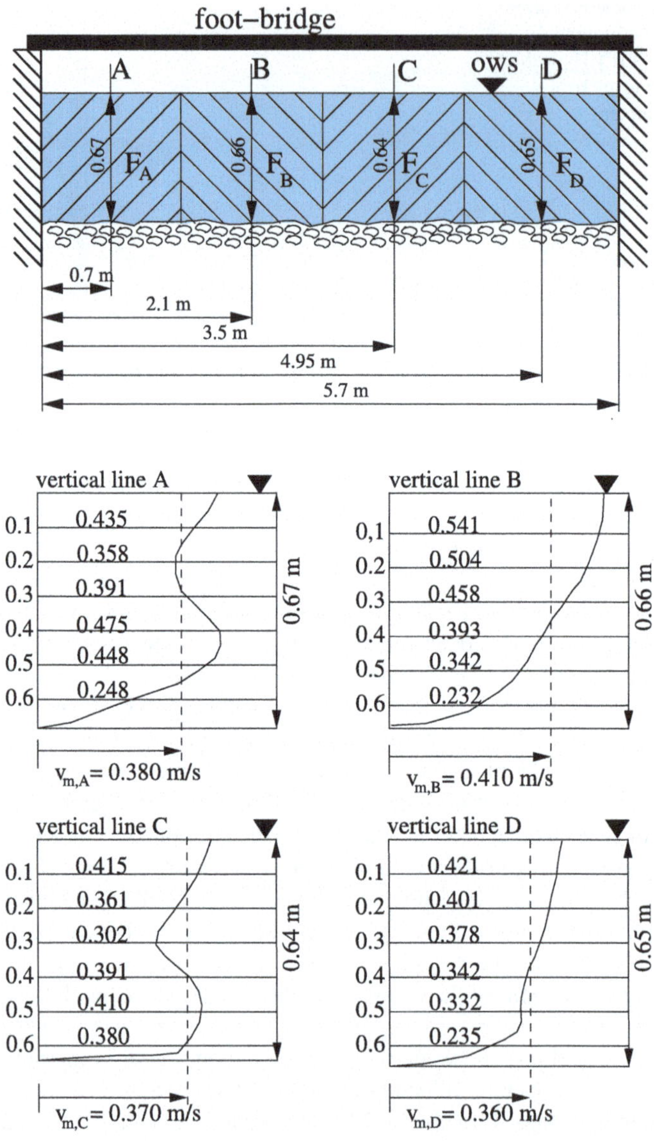

Fig. 9.8 Cross-sectional area and associated velocity distribution. Data are taken from [7]

$$c_{mC} = 0.37 \, \frac{m}{s} \quad F_C = 0.64 \cdot 1.40 = 0.896 \text{ m}^2$$

$$c_{mD} = 0.36 \, \frac{m}{s} \quad F_D = 0.65 \cdot 1.50 = 0.975 \text{ m}^2$$

The discharge was thus:

$$Q = \sum F_i \cdot c_{m,i} \quad \text{with} \quad i = A \ldots C$$

$$Q = 1.42 \, \frac{\text{m}^3}{\text{s}}$$

Errors of less than 2% can be achieved with this measurement method if the measurement is carried out with great care. This requires no mis-alignment of the hydrometric current meter to the flow direction, steady flow conditions and measurement of mean velocity values over a longer time interval.

9.3.7 Measurement of Velocity by a Float

The simplest method is the velocity measurement using a surface float. This can be a hollow ball, a bottle, or the like. It is important that the float submerges deep in the water to avoid influence by air resistance at the water surface (wind). The flow path should be as straight as possible. Two cross profiles are determined at a sufficiently large distance to accurately measure the time taken by the float between the cross sections.

It should take the float at least 30 seconds to travel between the cross-sections. The float will be handed over to the water current line about 20 m before the first profile. An observer determines the passage through the cross-section upstream, and likewise, a second observer determines the passage downstream. From the elapsed time and the path length, one can determine the water surface speed.

The experiments are to be repeated several times, and the average is to be formed. Float measurements are inaccurate because only the averaged surface speed is determined. The speed at the bed of the stream can be substantially lower. A detailed description of the flow measurement in open channels with various types of floats and the associated evaluation can be found in Herschy [16].

9.3.8 Flow Measurements Using Rising Air Bubbles

According to Sargent [22] and Herschy [16] one can perform flow measurements using rising air bubbles, which are used as floats. This method uses the behavior of air bubbles rising in water, which have a nearly constant ascent speed [22].

Assume a section from a channel with a rectangular cross-section and constant flow conditions, as shown in Fig. 9.9. The path of an air bubble from the bottom of the channel of water depth h to the surface and the velocity profile are shown in Fig. 9.9.

For this, the air bubble needs a time of $t_a = \frac{h}{v_z}$, where v_z is the constant ascent speed of the air bubble in the z-direction. At the same time, however, the air bubble drifts a distance L_i in the x-direction. This is due to the flow velocity v_x, which varies over the water depth. On average, a velocity v_m is established. The time is given by $t_a = \frac{L}{v_m}$. By eliminating the time from the relations, one obtains:

$$L \cdot v_z = h \cdot v_m$$

The product of $L \cdot v_z$ is equal to the product $h \cdot v_m$. The flow for a rectangular cross section A is determined with the following equation:

$$Q = A \cdot v_m = b \cdot h \cdot v_m = b \cdot L \cdot v_z$$

The ascent speed of an air bubble was determined in experiments by Sargent [22] on average to be $v_z = 0.218$ m/s .

The method's distinctive feature is that it does not require precise knowledge of the flow velocity or water depth h. It is sufficient to lay down a hose with nozzles in a riverbed, from which air bubbles escape perpendicular to the flow direction. Then the distance L has to be captured, which is the distance of the boundary line of the bubble veil at the water surface to the upstream hose line.

The advantage of the method is that even the nozzles can have a variable distance b from each other. But it is sometimes difficult to measure the length L very precisely (non-vortex free flow, parallaxes)

Now, however, one can divide a non-rectangular channel approximately into many rectangles. In this case, one determines discrete pairs of values of b_j and L_j. If one now divides a channel into sections perpendicular to the flow direction, one can determine the volume flow q_i flowing through the section in the following way $q_i = w_i \cdot v_z \cdot \frac{L_i + L_{i+1}}{2}$, where the length $\tilde{L}_i$ has already been averaged between the two measurement points $\tilde{L}_i = \frac{L_i + L_{i+1}}{2}$ and $w_i = (b_{i+1} - b_i)$. In Fig. 9.9, a partial area from a channel is shown. The partial flow is crossing the area $A = h_i \cdot w_i$ (see Fig. 9.9). The total flow then results from the sum of the individual flows

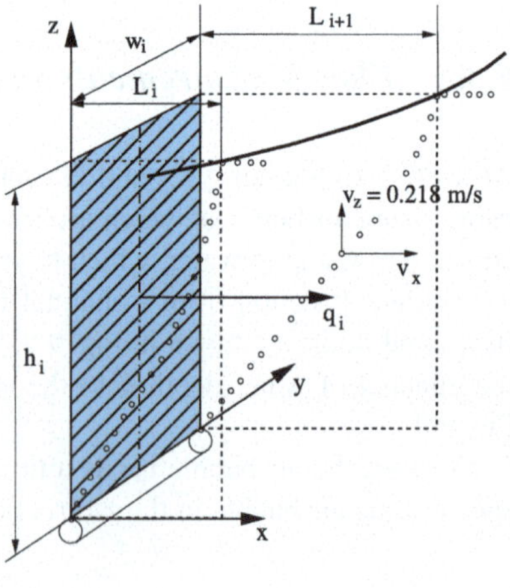

Fig. 9.9 Determination of flow with rising air bubbles in a section of the area $A = h \cdot b$

$q_i = w_i \cdot \tilde{L}_i \cdot v_z.$

$$Q = \sum_{j=1}^{n} q_j = v_z \cdot \sum_{i=1}^{n} w_i \cdot \tilde{L}_i$$

If the channel is measured as in Fig. 9.9, the following equation is used:

$$Q = v_z \cdot \sum_{i=1}^{n} (b_{i+1} - b_i) \cdot \frac{L_i + L_{i+1}}{2}$$

The accuracy of the method is comparable to that of hydrometric current flow measurements according to Herschy [16]. According to Sargent [22], an accuracy of – 6%...10% is achievable which is often sufficient for hydropower site planning purposes.

9.3.9 *Determination of the Discharge from a Precipitation Area*

Every river is formed by the confluence of several streams, each of which has its own special precipitation catchment area. The rain falling on an area is supplied to the river as groundwater or surface water. The individual catchment areas are separated from each other by watersheds, where the water drains off on both sides.

By knowing the total precipitation area in square meters, the amount of water flowing out of the area can be approximately determined as an annual average if the amount of precipitation in millimeters of water height per square meter is known. One millimeter of water height corresponds to one liter of water per square meter.

It can be concluded that the amount of rainfall contributing to discharge in larger rivers can reach up to 20...30%. In smaller watercourses and streams, the proportion of annual water quantity that contributes to discharge can reach 40...70%, which is considerably larger.

Thus, the average discharge of a given area as an annual average can be determined with the equation:

$$Q = \frac{\sum h_{NS} \cdot A_{NSG}}{3.1536 \cdot 10^7} \quad \text{in} \quad \left[\frac{m^3}{s}\right]$$

since the year has $t = 3.1536 \cdot 10^7$ seconds. A_{NSG} is the size of the precipitation area in $[m^2]$ and h_{NS} the precipitation height in $[m]$. The summation symbol is used when areas with different precipitation heights h_{NS} have to be considered. It is advantageous if precipitation maps are available for an area.

An annual precipitation amount of $h_{NS} = 800$ mm would cause an average discharge of $Q = (0.4\ldots0.7) \cdot \frac{0.8\ \text{m}\cdot1\cdot10^6\ \text{m}^2}{3.1536\cdot10^7\ \text{s}} = 0.010\ldots0.0178\ \text{m}^3/\text{s} =$ $10\ldots17.8$ l/s per square kilometer (assuming a $40\ldots70\%$ discharge rate).

9.4 Measurement of the Rotational Speed

The speed of rotation can be measured with mechanical tachometers. However, today the speed of rotation is often converted into electrical signals to further electronic control or regulation devices to provide this important system information. In general, pulses are measured and compared with a time standard. The speed can be determined from the number of pulses per time interval. Methods for speed measurement include, for example, the optical method or inductive methods. Speed measurement is also necessary for control when starting and stopping the waterwheel.

9.4.1 Optical Measurement of Speed

The principle of rotational speed measurement is shown in Fig. 9.10. A light source illuminates a disc mounted on a shaft. This ring disc has black and white stripes. A phototransistor is used as a detector, which emits an electrical signal depending on whether a black circular segment or a white one passes by. In the same way, one can use perforated discs. The white stripes correspond to a logical one and the black ones to a logical zero. This results in a square-wave pulse sequence, which is proportional to the speed.

This pulse sequence must be compared with a time standard. For this purpose, for example, a pulse counter is used, which is reset after a certain time interval (e.g. 1 s, 1 min). The final result of the count is stored temporarily and passed on to an evaluation circuit (digital display, control etc.). Another possibility is the use of perforated discs instead of black and white stripes.

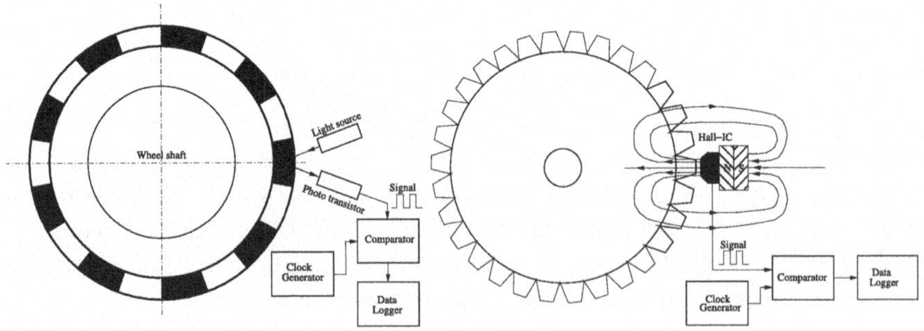

Fig. 9.10 (**a**) Optical, and (**b**) Inductive rotational speed measurement

9.4.2 Measurement of Speed with a Hall Effect Sensor

In Fig. 9.10b, the operating principle of this measuring device is shown. The integrated circuit (IC) contains a Hall plate, which is sensitive to magnetic fields. A Hall plate provides a Hall voltage proportional to the magnetic field, which is amplified by an integrated voltage amplifier. Usually, there is also a digital evaluation logic in the circuit, which can detect the minima and maxima of the magnetic field.

As a tooth of a ferromagnetic gear wheel passes by, the magnetic field changes, as the air gap decreases. The Hall IC provides a square pulse for each passage of a tooth. The square pulses are counted and output via a resettable counter. The pulse count is then transferred to the rotational speed knowing the diameter of the disk with the teeth.

9.5 Measurement of Torque

The torque is determined with brake dynamometers. These are power brakes, with which the power is determined by measuring the braking force. The braked power is consumed and converted into heat.

In the **mechanical brake dynamometer**, brake pads or a brake belt (Prony brake) are pressed against a brake disc (see e.g. Fig. 9.11). The braking torque is balanced by a weight on a lever arm and measured with a scale. The Prony brake was often used in waterwheel measurements in the past, including by Prof. Staus [5, 17] Meerwarth [6], W. Müller [23]. A very good overview and hints for the application of the Prony brake can be found at Staus [17] However, the method is

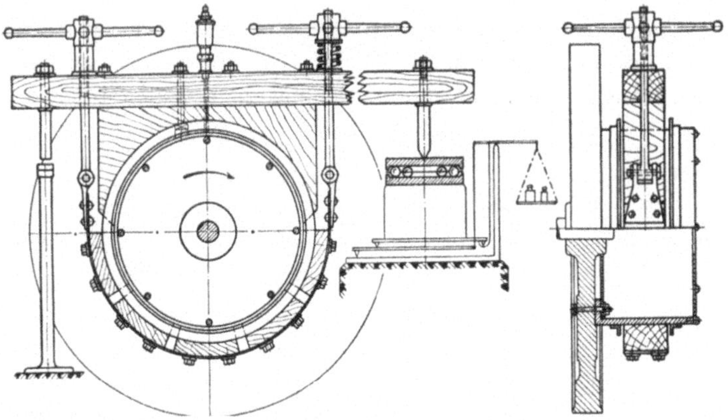

Fig. 9.11 Realization of a Prony brake for a 3 kW combustion engine. The Figure was taken from Staus "Maschinenuntersuchungen", Julius Springer Verlag 1926 [17]

problematic for larger powers due to the heat development (water cooling becomes necessary).

In the **hydraulic brake dynamometer** (water whirl brake), the internal friction of the fluids during shape changes is exploited. The frictional forces increase with the square of the shape change speed, so the water whirl brake is particularly suitable for high speeds. The **electric brake dynamometers** are eddy current brakes, in which a disc rotates in a finely adjustable magnetic field. This induces eddy currents in the disc which generate forces or moments that oppose the generating motion (Lenz's law). This reaction moment is made measurable by pendulum-mounted brake magnets.

For large torques, **torsion dynamometers** are used, which operate without power consumption. A torsion bar, which is clamped between two shaft ends, serves as a sensor. There is a functional relationship between its twist (in the elastic range), which is measured, and the torque. The electrical measurement of the deformation is based on the resistance change of an electrical conductor during length changes. For this purpose, strain gauges are used, which are shear-resistant and glued to the shaft at an offset of 45°. The reporting of the electrical signal is done via slip rings or contactless rotary transmitters (radio).

The **back pressure method** uses the reaction moment for measurement, which is otherwise absorbed by the machine foundation. For this purpose, the drive machine or the brake is pendulum-mounted. This method is often used for power measurement of electric and combustion engines and can be permanently installed.

The Prony brake will be explained in more detail because it was used often at large scale experiments on waterwheels in the past. This knowledge is needed in the last chapter of the book.

9.5.1 *Mechanical Brake Dynamometer (Prony Brake)*

The Prony brakes or Prony belt brakes were often used in the past to measure power of rotating machinery [6, 13, 17, 23]. The brake works as follows (see Fig. 9.12). A brake disc is mounted on an output shaft of the mechanical system

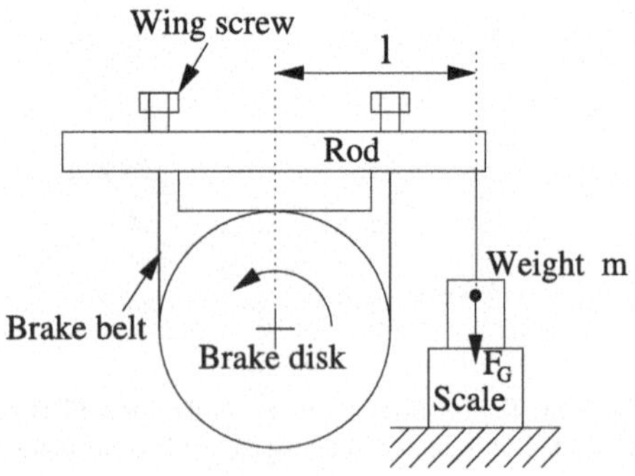

Fig. 9.12 Prony brake

under investigation. The braking force, acting on the rotating shaft by the brake pads, can be adjusted using the brake screws.

On top of the brake disk a rod is screwed in, on which a weight is hung at position l. By tightening the brake screws, the friction between brake pad and shaft is increased, thus creating a counter moment on the brake pad. The rod is kept in balance by the weight hanging on it.

As soon as the rod begins to lift off the support, the equilibrium state has been reached, and the torque of the wheel is equal to the counter moment of the brake. The lifted weight is then determined by a weighing scale. The torque on the shaft is given by:

$$M = r \cdot F_r$$

where r is the radius of the brake disc and F_r is the acting friction force. This torque is equal to the counter moment, which is determined by the rod length and the attached weight

$$M = l \cdot F_g = l \cdot m \cdot g$$

The power output of the waterwheel is calculated with the equation:

$$P = M \cdot \omega = M \cdot \frac{2 \cdot \pi \cdot n}{60}$$

The rotational speed is to be given in the equation in $[min^{-1}]$. By substituting the above formulas, one obtains the very simple relationship:

$$P = M \cdot \omega = l \cdot m \cdot g \cdot \frac{2 \cdot \pi \cdot n}{60}$$

So, one has the mass m, the length[2] l and the speed n to determine the power output of the wheel. It should be noted that the Prony brake is completely balanced, that is, the brake must be symmetrically built to the center of the shaft of the countershaft.

[2] Staus [5] and Meerwarth [6] use a rod length in their experiments from $l = 0.716$ m, to be able to calculate the power directly in horsepower over the weight change and speed, since the following applies:

$$P = l \cdot m \cdot g \cdot \frac{2 \cdot \pi \cdot n}{60 \cdot 0.7355} = m \cdot n \quad \text{in [hp] horsepower, since}$$

$$0.716 \cdot 9.81 \cdot \frac{2 \cdot \pi}{60 \cdot 0.7355} \approx 1$$

To obtain information in the same way in kilowatts, one would choose a rod length of $l = 0.973$ m.

Since the power is converted into friction heat, the brake disc must be cooled by water. Large mechanical powers are therefore not measurable with this method (Limit ca. 100 kW). Another disadvantage is the high mechanical effort for such a measurement. A direct measurement on the waterwheel shaft is excluded due to the high torques.

Usually, mounting a disk brake on a shaft in today's closed gearboxes is no longer possible. As a result, attaching a Prony brake to a modern waterwheel system is challenging.

In addition, the transmission efficiency η_{gear} of the countershaft (e.g. in spur gear drives) under load is not known accurately enough. To stabilize the braking, an additional flywheel is necessary. A discussion on this problem can be found in Meerwarth [4, 6]. The method is therefore rarely used (e.g. in laboratory experiments [9, 24]).

9.6 Measurement of Electrical Power

In the single-phase alternating current (AC) electrical network, the average of the power (active power) is measured with the wattmeter. The average of the power results from the current I through the wattmeter and the voltage across the wattmeter.

If one wants to measure the active power in each phase of the three-phase system, a wattmeter is switched into each strand of the electrical network. The total power is the sum of the individual values displayed at each wattmeter. The use of only one wattmeter is sufficient if the symmetric electrical network of an asynchronous generator is measured, whereby the displayed value of the wattmeter must be tripled.

The circuit variants shown in Fig. 9.13 are used if there is no star point wire. Connection assignments for wattmeter can be found in Friedrich [25].

For large powers $P_{elec} \geq 10$ kW, commercial wattmeters are unsuitable. Under these circumstances, measuring transformers are used, or the currents in the supply lines of the generator are measured using a current clamp. With current clamps,

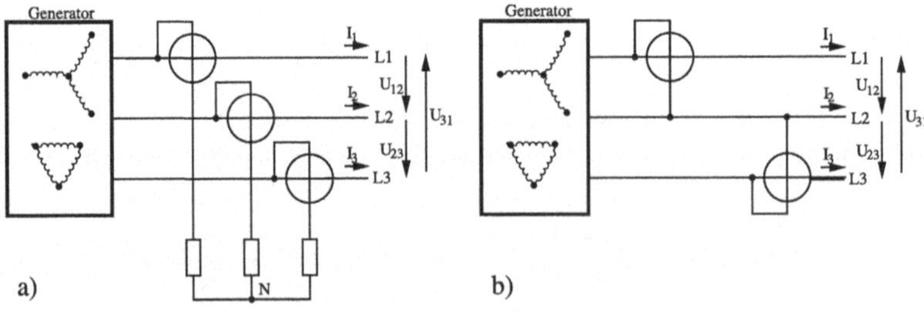

Fig. 9.13 (**a**) Power measurement without neutral conductor. (**b**) Aron circuit

currents up to 1000 A can be measured. At the same time, the voltage between the different wires must be measured with a voltmeter (multimeter).

The power factor of the electrical machine must be taken into account, as a phase shift between current and voltage may be present. The power factor is strongly dependent on the load conditions of the electrical machine. One has to remember that the power factor is only specified for the maximum power on the type plate of the electrical machine.

Therefore, it is better to determine the phase shift between the current and the voltage with an oscilloscope. To check the symmetry of the three-wire system, it is also useful to measure the voltages relative to the ground point.

Another method for determining electrical power is the speed measurement at the generator. The electrical power is almost linearly dependent on the generator speed. The implementation and evaluation of such a measurement can be found in Chap. 10.

Typically, the operator of a system does not have access to a wattmeter, current clamp, multimeter, or oscilloscope. A simple and yet very accurate method is to measure the speed of the kilowatt-hour meter, which is always present when feeding electricity into the grid, with a stopwatch.

In parallel grid operation, of course, all electrical consumers in the local network must be switched off. The meters are calibrated to less than 1% error, and the meter indicates the revolutions per kilowatt-hour. From the time measurement, one can determine how many revolutions the meter would have made in an hour and thus calculate how much power is delivered by the electrical machine per second.

The particular advantage of this simple measurement method is that the measurement period can be chosen arbitrarily long (integration time) and that load fluctuations due to impacts or imbalance of the wheel are averaged out.

Example 9.4. ***(Determination of Electrical Power with a Meter Measurement)*** **The following Table 9.3 summarizes the results of a meter measurement. In the second column, the time for 10 revolutions of the meter can be found and in the third column the electrical power reading on the wattmeter is given. The meter indicates that one kilowatt-hour corresponds to 75 revolutions. The calculated electrical power from the meter measurement times (column 2) is to be compared with the readings of the wattmeter installed in the system (column 3) and the measurement error is to be determined.**

Table 9.3 Data of power measurements

No	t [s]	$P_{elec,wattmeter}$ [kW]	$P_{elec,kWh-meter}$ [kW]	Measurement error [%]
1	50	8.3	9.6	–13.5
2	69.5	6.7	6.9	–2.9
3	98.0	4.4	4.9	–10.2
4	146	2.9	3.3	–12.1
5	52.3	8.6	9.2	–6.5
6	45.95	9.7	10.4	–6.7

The meter measures the electrical work of the system in kilowatt-hours. The electrical power is then given by $P = \frac{W}{t}$. The meter indicates that one kWh corresponds to 75 revolutions. If the time for 10 revolutions is measured, the following equation results (1 kWh = 3600 kWs):

$$P_{el,kWh-meter} = \frac{3600\,\text{kWs} \cdot 10\,\text{U}}{t \cdot 75\,\text{U}} = \frac{3600\,\text{kWs} \cdot 10\,U}{50\,\text{s} \cdot 75\,\text{U}} = 9.6\,\text{kW}$$

t is the time that needs to be measured. The calculated values are given in Table 9.3. The measurement error of the wattmeter results from:

$$\epsilon = \left(\frac{P_{el,wattmeter}}{P_{el,kWh-meter}} - 1\right) \cdot 100\% = \left(\frac{8.3\,\text{kW}}{9.6\,\text{kW}} - 1\right) \cdot 100\% = -13.5\,\%$$

In Table 9.3, all calculated values are given in the fourth column. The measurement error of the wattmeter can be found in column 5. During this measurement, it was found that the wattmeter in the switch cabinet was faulty. The device showed power values that were 3 ... 13.5% too low. The wattmeter was no longer usable. The device was replaced.

References

1. Smeaton J (1759) An experimental enquiry concerning the natural powers of water and wind to turn mills, and other machines, depending on a circular motion. Philos Trans R Soc 100–174
2. Zeuner G (1855) Über ein eisernes Wasserrad mit Coulissenschütze zum Betriebe der Schneidemühle in Deuben bei Dresden. Polytech J 14:834–838
3. Müller W (1911) Anlage zweier Zuppinger-Niedergefälleräder von 40 bis 100 PS Leistung. Polytech J 326:774–777
4. Weidner CR (1913) Theory and test of an overshot water wheel. Bullet Wisconsin Univ Eng Ser Madison Wisconsin 7(529):157–254
5. Staus A (1928) Wasserradversuche. Wasserkraft und Wasserwirtschaft 18:259–260
6. Meerwarth KW (1935) Experimentelle und theoretische Untersuchungen am oberschlächtigen Wasserrad. PhD Thesis, Technische Hochschule Stuttgart
7. Neumayer H, Rempp W, Ruppert J, Schwörer R (1977) Untersuchungen am Wasserradtriebwerk der Kunstmühle W. Seifried KG, Waldkirch/Br
8. Günther T (1997) Das Turas-Wasserrad, ein einseitig gelagertes Wasserrad. Wasserkraft und Energie, Verlag Moritz Schäfer 2/97:52–56
9. Müller G, Wolters C (2004) The breast shot water wheel: design and model tests. Proc ICE Eng Sustain 157(ES4):203–212
10. Dubas M (2005) Design and efficiency of overshoot water-wheels. Int J Hydropower Dams 12(3):359–368
11. Nuernbergk DM, Trostdorf F, Treiber G (2009) Untersuchung zum Wirkungsgrad eines Zuppinger-Wasserrades mit einer kreisförmigen Überfallschütze. In: Wasserbaukolloquium 2009, pp 359–368
12. Nuernbergk DM, Trostdorf F (2010) Effizienzsteigerung in Energieerzeugungsanlagen durch Stern-Dreieck-Umschaltung. Wasserkraft und Energie 1:53–69
13. Müller W (1999) Die Wasserräder – Berechnung, Konstruktion und Wirkungsgrad, 2nd edn. Verlag Moritz Schäfer, Detmold
14. Knauer H (1923) Der Wasserbau. Polytechnische Verlagsgesellschaft Max Hittenkofer, Strelitz in Mecklenburg
15. Bureau of Reclamation – U.S. Department of the Interior (2001). Water measurement manual: a water resources technical publication – a guide to effective water measurement practices for better water management. https://www.usbr.gov/pmts/hydraulics_lab/pubs/wmm/

16. Herschy RW (1999) Streamflow measurements, 2nd edn. Chapman & Hall, London
17. Staus A (1923) Maschinenuntersuchungen - I. Band: Hydraulik, 2nd edn. Verlag von Julius Springer
18. Henne, H (1903) Die Wasserräder und Turbinen – ihre Berechnung und Konstruktion. Verlag von Bernh. Friedr. Voigt, Leipzig
19. Bollrich G, Preißler G (1992) Technische Hydromechanik Band 1, 3rd edn. Verlag für Bauwesen, Berlin
20. Wahl TL, Clemmens AJ, Bos MG, Replogle JA (2013) Winflume, U.S. Department of the Interior, Bureau of Reclamation. www.usbr.gov/pmts/hydraulics_lab/winflume/
21. Woltman R (1790) Theorie und Gebrauch des Hydrometrischen Flügels oder eine zuverlässige Methode die Geschwindigkeit der Winde und strömenden Gewässer zu beobachten. Theory and use of the current meter or a reliable method of observing the velocity of wind and streams. Verlag von Benjamin Gottlob Hoffmann, Hamburg
22. Sargent DM (1981) The development of a viable method of stream flow measurement using the integrating float technique. Proc. Instn Civ. Eng. 71:1–15
23. Müller W (1899) Die eisernen Wasserräder – Berechnung, Konstruktion und Bestimmung ihres Wirkungsgrades, erster Teil – Die Zellenräder, 1st edn. Verlag Veit und Comp.
24. Becker T (2004) Physikalische Modellversuche am mittelschlächtigen Wasserrad zur Optimierung der Geometrie und des Wirkungsgrads. Diplomarbeit, Dezember
25. Friedrich (1998) Tabellenbuch Elektrotechnik Elektronik. Dümmler Verlag Köln

Chapter 10
Experimental Investigations

10.1 Overshot Waterwheel Test Under Laboratory Conditions

10.1.1 Description of the Experimental Setup

At this point, Meerwarth [1] should be quoted verbatim: The wheel, a standard design by Fa. Müller, Bad Cannstadt, has a diameter of $D = 3.6$ m measured for the cell tips, a wheel width of $B = 0.8$ m, a cell depth of $a = 0.28$ m and $z = 30$ steel sheet cells. It is designed for a design flow rate of $Q = 0.15\ \text{m}^3/\text{s}$ and a head difference of $H = 4.0$ m (Fig. 10.1).

The normal rotational speed is $n = 9\ \text{min}^{-1}$ at the wheel shaft. To achieve reasonable dimensions for the brake, the waterwheel was equipped with a double gear set with a transmission ratio of 1 : 25. The brake gear wheel therefore runs at 25 times the speed of the wheel.[1]

The flywheel, with a diameter of 800 mm and a mass of 245 kg, is keyed to the shaft of the brake gear and significantly enhances braking stability, particularly at low speeds. All shaft journals run on roller bearings. All spur gears are milled.

The inlet device consists of an inlet box, in which two calming grids made of wood are connected in series. The operating water from the centrifugal pump is supplied to it from below via the throttle slide.

Three circular overflow weirs with a clear width of 275 mm are built into its front vertical wall. The overflow weirs are separated by partitions. With these partitions, the exact conditions under which the overflows were calibrated in the experimental channel are obtained. The overfall height is read off a float level with a parallax-free

[1] Staus gives the following tooth numbers for the gear set: 1. gear set 163/32, 2. gear set 150/30. The resulting transmission ratio is $i = 1 : 25.47$. However, the transmission ratio $i = 1 : 25$ is used in the following calculations.

D. M. Nuernbergk, *Overshot Waterwheels*, Springer Tracts in Mechanical Engineering, https://doi.org/10.1007/978-3-031-95385-9_10

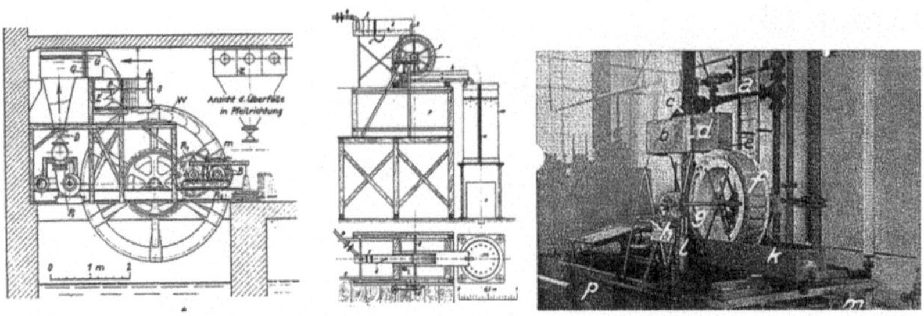

Fig. 10.1 Experimental setups by Meerwarth in the hydraulic laboratory in Esslingen [1], Left: Schematic representation of the large wheel, Middle: Schematic representation of the model waterwheel, Right: Photo of the experimental setup with the model wheel. The meanings of the letters in the photo are: a—feed line with throttle valve, b—feed channel, c—calming grid, d—glass level with scale, e—control contactors, f—model wheel, h—Prony belt brake, k—tailwater channel, l—float level, m—hole danaide

millimeter scale and simultaneously on a standpipe. Each overflow is connected to the float level via a rubber hose.

From the overfall weirs, the water, guided by the baffle plates, falls into the sluice box, which ends in a "nozzle" that is 600 mm wide. The closing sluice has a fine adjustment with a millimeter scale and vernier.

Up to a load of $Q = 0.18$ m^3/s, the centrifugal pump from the test setup is sufficient. The flow rate it takes from the measuring channel is measured directly with the three circular overfall weirs. From $Q = 0.18$ m^3/s onwards up to the overload limit of the wheel, the second pump must be used. This pump also draws its water from the flow measuring channel, but pushes it through a special pipe directly into the shutter box.

The separate supply is necessary because a parallel connection of the two pumps, as was once attempted, does not increase the flow rate beyond $Q = 0.18$ m^3/s. As a result, one must measure the flow of the two pumps together with the measuring screen running in the measuring channel.

The calibrated flow rate measuring channel is 1.5 m wide and has a usable measuring length of 9 m. It is specifically set up for screen measurements[2] and therefore has sliding contacts at intervals of 1 m which are triggered by the screen. The running times from the screen are recorded by a time writer. At a float level with millimeter scale, the water levels in the measuring channel are read off.

Now, however, a test turbine is connected to the measuring channel. During the tests, therefore, the leakage water flow of this turbine had to be determined by special tests and subtracted from the water flow to the wheel.

[2] A screen attached perpendicular to the measurement channel is used to measure the velocity of the flowing water. If the cross section is known, the flow rate can be determined very accurately. Screen measurements are described in the book of Staus [2].

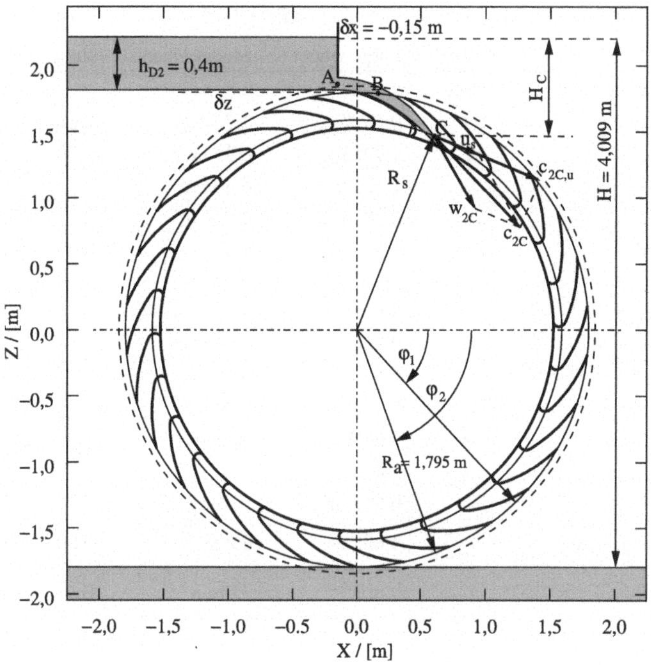

Fig. 10.2 Schematic setup of the experimental waterwheel for $H = 4$ m. The wheel dimensions were as follows: diameter $2 \cdot R_a = 3.592$ m, wheel width $B = 0.8$ m, and rim width $a = 0.27$ m, Number of cells $z = 30$, pitch $t_z = 0.376$ m. The incoming water jet was drawn for a flow rate of $Q = 0.148\ \mathrm{m^3/s}$

Since the wheel tends to splash a lot at higher speeds and with large impingement, it is covered for the protection of the observers by splash guards, which only have openings at the water inlet and at eye level through which the processes in the wheel can be observed.

The overshot waterwheel used in the experiments is shown in Fig. 10.2 to scale. The dimensions of the large wheel are given in Table 10.1.

In addition to the overshot waterwheel test facility by Prof. A. Staus, Meerwarth designed a smaller model wheel on a 1:5 scale to investigate the influence of the immersion depth H_T on the efficiency of the wheel because it was not possible to test the wading of the large wheel.

The model wheel will not be considered further. However, it is of interest that it had the same efficiencies as the large wheel. For those interested, the excellent PhD thesis by Meerwarth [1] is recommended. The details of this scaled waterwheel can also be found in Table 10.1.

Table 10.1 Dimensions of the overshot large and model wheel

	Size	Large wheel	Model wheel	Remark
Head difference	H	4.0 m	0.8 m	
Flow rate	Q	0.15 $\frac{m^3}{s}$	2.7 [l/s]	Design flow rate
Outer radius	R_a	1.795 m	0.358 m	
Inner radius	R_i	1.525 m	0.304 m	
Rim width	a	0.28 m	0.054 m	
Filling factor	ε	0.3	0.3	For design-Q
Pitch	t_z	0.376 m	0.075 m	
Number of cells	z	30	30	
Wheel width	B	0.8 m	0.1604 m	
Channel width	b_0	0.6 m	0.12 m	
Pressure height	h_{D2}	0.4 m	0.08 m	
Immersion depth	H_T	−0.05 m	Variable	
Distance of the sluice gate	δx	−0.15 m		
to the apex	δz	0.017 m		
Circumferential speed	u_A	1.692 $\frac{m}{s}$	0.825 $\frac{m}{s}$	At the outer radius
Speed	n	9 min^{-1}	22 min^{-1}	Best speed

10.1.2 Experiments on the Large Wheel

The experiments were conducted with a Prony brake to determine the torque. Braking was consistently done at a constant flow rate Q, i.e. the experiments were carried out at a constant opening of the sluice gate s, at a steady pressure height h_{D2} in the shutter box, and at variable rotational speed. The flow measurements are based on the calibration curves of Staus [3]. The error was estimated to be less than 0.3% for all flows.

The useful head difference H was defined as the vertical distance between the upper water level and the lower wheel vertex of the cell tips. With $\delta z = 0.017$ m, $D_a = 3.592$ m it follows:

$$H = h_{D2} + \delta_z + D_a = h_{D2} + 3.609 \text{ m} = 4.009 \text{ m}$$

The rotational speeds were read off a speedometer mounted at the brake gear. In the case of no-load tests, a stabbing tachometer had to be used because the high speeds occurring could no longer be read accurately with the speedometer. The room temperatures averaged 9 °C during the tests. This corresponds to a water density of $\varrho = 999.7 \text{ kg/m}^3$.

The brake loads were balanced with a carefully balanced decimal scale for a maximum load of 100 kg. The brake, whose lever length was 716 mm, was completely balanced. The power was determined for flow rates of Q = 30, 60, 90, 120, 150, 180, 200, 210 l/s. Each series of experiments generally consisted of

Table 10.2 Excerpt from the experimental results of Meerwarth [1] (The conversion factor from kilowatt to hp (horsepower) is 0.7355)

Nr	$n \cdot 25$ [1/min]	m [kg]	h_{D2} [m]	H [m]	s_0 [m]	Q [l/s]	P_{th} [hp]	P_N [hp]	η [1]
61	304.0	26.5	0.390	4.000	0.0126	180.3	9.62	8.06	0.837
46	262.0	30.5	0.4	4.009	0.0126	181.5	9.70	7.92	0.82
43	221.0	36.7	0.399	4.008	0.0126	181.3	9.68	8.1	0.837
36	201.6	40.0	0.401	4.010	0.0125	179.9	9.60	8.06	0.840
57	181.0	44.5	0.4	4.009	0.0126	180.7	9.66	8.06	0.834
44	165.0	46.2	0.399	4.008	0.0126	181.3	9.68	7.60	0.785
45	153.0	47.3	0.4	4.009	0.0126	181.1	9.64	7.24	0.752
59	141.0	49.0	0.4	4.009	0.0126	181.0	9.67	6.90	0.714

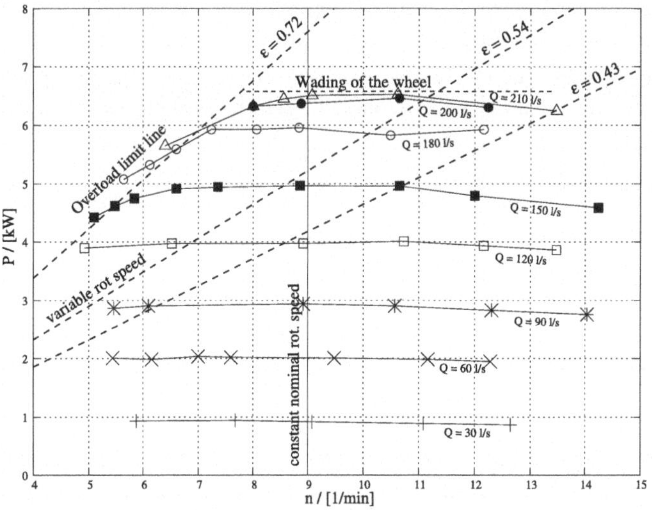

Fig. 10.3 Measured mechanical output power of the large overshot waterwheel

seven to ten test points. Each time, the process began at an average rotational speed of approximately $n = 200\ \text{min}^{-1}$ (waterwheel speed $n = 8\ \text{min}^{-1}$), measured at the brake gear. The speed was gradually reduced as far as the brake's inertia permitted.

However, at speeds below approximately $n = 120\ \text{min}^{-1}$ (corresponding to $n = 4.8\ \text{min}^{-1}$ for the waterwheel), maintaining the speed proved impossible despite careful monitoring. The high speeds posed no problems.

An excerpt from Meerwarth's [1] measurement results are compiled in Table 10.2. Since the lever length of the Prony brake was 716 mm, the second (rotational speed) and third (mass) columns in Table 10.2 are to be multiplied together to obtain the useful power in HP (horsepower).

The experimental results of the measured mechanical power are shown in Fig. 10.3. The data given in horse power in e.g. Table 10.2 were converted into kilowatts. The power is almost constant for different rotational speeds, if the wheel is not overloaded by wading, overspeed or overfilling. Only at very high rotational speeds higher than $n = 11 1/\text{min}$ the power curves bend down.

In this graph it can be seen that overfilling of the cells limits the power output of the wheel. The filling factor in this case is $\varepsilon = 0.72$. The output power for different constant cell filling factors are shown in Fig. 10.3 as straight lines.

The experimental results of the efficiencies are presented in Figs. 10.9 and 10.10 together with the results of the theoretical calculations.

The difference between the results for the efficiencies of Staus [3] and Meerwarth [1] arises from the gross head difference, which Staus states as $H = 4.063$ m and Meerwarth on average as $H = 4.009$ m. According to his own statements, Staus has considered a free hanging of the waterwheel of $H_T = -0.05$ m, while Meerwarth has measured the gross head difference from the upper water level to the lower wheel vertex of the wheel (edges of the cells).

It is important to note that the accuracy of the flow measurement is in the range of 0.3%. All results which are shown, are converted to the head difference given by Meerwarth. Meerwarth [1] will be quoted again to explain the experiments in the next subsection.

Experiments at an Inlet Height of $h_{D2} = 0.4$ m

To gain further insight into the operating characteristics of the waterwheel and to make comparisons with turbines, Fig. 10.10 shows the efficiency of the examined large wheel as a function of the flow rate. The waterwheel efficiency values apply to the best rotational speed of $n_n = 9$ min^{-1} and are based on the power at the wheel shaft.

The following can be concluded from the Figures: The efficiency of the waterwheel, plotted as a function of the load, is very flat. The maximum value of around $\eta = 0.85$ is only surpassed by small turbines of the best design. It should be noted that the examined waterwheel ran with two gearboxes in all experiments. Without these two gearboxes, its maximum efficiency value would be even higher.

Up to a load of around $Q/Q_{max} = 0.65$, the efficiency of the waterwheel is significantly above the efficiency of the small turbine. The no-load water flow is $Q = 0.0033$ m^3/s (3.3 l/s) to turn the wheel at $n_n = 9$ min^{-1}. This is around 1.6% of the maximum water flow. In comparison the no-load water flow in the small turbines do not drop below 7% of the maximum water flow in the best case. Staus [3] has already pointed out this advantage.

At $Q = 0.21$ m^3/s, the waterwheel becomes overloaded. This flow is 40% above the design flow of $Q = 0.15$ m^3/s.

Furthermore, Figs. 10.9 and 10.10 show how little the efficiency changes when deviating from the best rotational speed. Within a discharge load of around 0.3…0.7, the efficiency remains above 80% between $n = 6$ and 14 min^{-1}. The overspeeds of the waterwheel are significantly higher than those of turbines. They reach approximately three times the normal speed.

It is also worth noting that the torque curves of the waterwheel, expressed as a function of speed $M = f(n)$, exhibit an approximately hyperbolic trend, in contrast to the torque curves of the turbine. The curves deviate significantly from this pattern at high flow rates when the speeds n become low, as the cell filling becomes excessively large. The cells are overfilled then.

As a result, the cells begin to empty prematurely in the entry range. The water is expelled from the cells and shoots out in an almost closed jet over the wheel. The contents of the cells are thoroughly swirled, though bubble-free. At high speeds, the jet widens progressively as it strikes the blade, initially causing a slight edge veil.

This edge veil increases in size with increasing speed and load that water splashes laterally over the wheel discs. Finally, it is a closed jet which shoots smoothly over the cells. This jet increasingly covers the cells and thereby prevents the air trapped in the cells from escaping. The torque of the tested wheel drops very quickly.

Experiments at an Inlet Pressure Height of $h_{D2} = 0.2$ m

Compared to the experiments with a $h_{D2} = 400$ mm inlet pressure height, the operating range has become slightly smaller here. The smaller inlet height of $h_{D2} = 200$ mm is noticeable in that the efficiency curves shift to the side towards lower speeds.

The efficiency remains practically unchanged. A moderate change in the inlet height therefore has no noticeable effect on the efficiency. Small water level fluctuations, as occur in water reservoirs (e.g. ponds or lakes), can be allowed without concern.[3]

10.1.3 Measurements Versus Theory

It was of great interest to recalculate the results of Staus [3] and Meerwarth [1], as these experiments were conducted very carefully and thus a direct comparison of the theoretical results of this work was expected. In this section the mechanical and hydraulic losses are calculated.

The comparison revealed a surprisingly good agreement. This could only be achieved by considering the deflection of the jet at the entry of the water into the cells. The deflection has not been accounted for in the efficiency calculations in the literature (e.g. [4, 5]), which is why lower efficiencies for overshot waterwheels were determined with the methods available at this time.

[3] This result is also important for a water supply via an open channel, as the speed at the outlet of the channel strongly depends on the water depth in the channel. A change in the inlet speed only very weakly changes the efficiency of the overshot waterwheel.

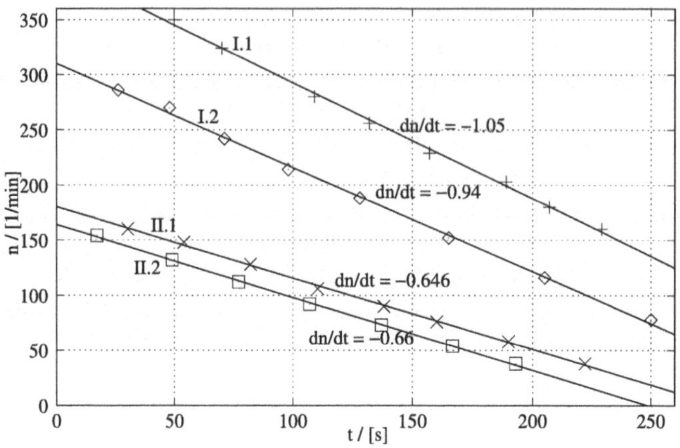

Fig. 10.4 Results of the run-out experiments [1]: I. Run-out experiments with both gears (used in the calculations), II. Run-out experiments with both gears, cleaned and filled with new grease

10.1.4 Mechanical Losses

First, the mechanical losses are determined. Meerwarth [1] conducted experiments with a free running, unloaded waterwheel (see Fig. 10.4). In this case, the wheel is accelerated at the beginning and then left to itself. The decrease in speed over time is a measure of the friction losses.

From these experiments, the mechanical losses can be accurately determined in retrospect. However, the moment of inertia J must be known for this calculation.

In the experiments, the moment of inertia was determined (calculation of the moment of inertia of the waterwheel and the 1st gear, torsional oscillations for determining the moment of inertia of the brake gear) and the following mass moments of inertia were given:

$$J_1 = 390 \text{ kg} \cdot \text{m} \cdot \text{s}^2 \quad \text{waterwheel (calculated)}$$

$$J_2 = 1.94 \text{ kg} \cdot \text{m} \cdot \text{s}^2 \quad \text{1st gear (calculated)}$$

$$J_3 = 1.745 \text{ kg} \cdot \text{m} \cdot \text{s}^2 \quad \text{2nd gear (measured)}$$

The data is not given in the usual unit $\text{kg} \cdot \text{m}^2$, since the SI units were not used at this time. The torque was denoted with $\text{m} \cdot \text{kg}$. Therefore, the given moments of inertia still have to be multiplied by the standard gravitational acceleration $g = 9.80665 \text{ m/s}^2$.

The total moment of inertia related to the brake mechanism is then:

$$J_{total} = J_3 + J_2 \cdot \left(\frac{1}{5}\right)^2 + J_1 \cdot \left(\frac{1}{5} \cdot \frac{1}{5}\right)^2 = 24.0 \text{ kg m}^2$$

From the run-out tests, we now take the drop in speed over time $(n - n_0)/(t_0 - t) = 0.94\ 1/(s \cdot \text{min})$ from Fig. 10.4 (Curves I.) and insert this into the equation for the friction torque (derivation see [6]).

This gives us the friction torque related to the brake mechanism:

$$M_{R,3} = \frac{J \cdot 2 \cdot \pi}{60} \cdot \frac{n(t_0) - n(t)}{t_0 - t} = \frac{24 \cdot 2 \cdot \pi}{60} \cdot 0.94 \frac{\text{kg m}^2}{\text{s}^2} = 2.36\ \text{Nm}$$

This still needs to be converted to the waterwheel shaft:

$$M_{R,1} = \frac{M_{R,3}}{i} = 2.36 \cdot 25\ \text{Nm} = 59.0\ \text{Nm}$$

Fortunately, Meerwarth has determined the losses of 0.077 hp for the best speed of $n_3 = 225\ \text{min}^{-1}$ at the Prony brake. A comparison shows:

$$P_R = M_{R,1} \cdot \frac{2 \cdot \pi \cdot n_1}{60} = 55.7\ \text{W}$$

With the conversion factor from kilowatts to horsepower hp of 0.7355, we get a power loss due to friction of 0.0757 hp. This result agrees well with the value given by Meerwarth.

The power loss due to friction is not constant, as is often assumed because it depends on the rotational speed. But the friction torque M_R is a constant. This also means that gearboxes do not have constant efficiencies since the efficiency depends, for example, on the rotational speed and the load.

The advantage of the precisely known mechanical losses in this case is obvious: One can separate the mechanical and hydraulic losses, and a precise recalculation of the results is possible. In Fig. 10.5, the mechanical losses are represented as loss heights:

$$h_{V,mech} = \frac{P_R}{g \cdot \varrho \cdot Q \cdot H}$$

plotted. One can clearly see the high percentage of mechanical losses at low flow rates. The efficiency of the gearbox is lower at a small flow rate than at larger ones.

For example, for a flow rate of $Q = 20$ l/s at the best speed of $n = 9\ \text{min}^{-1}$, about 10% of power is lost due to mechanical losses, while at $Q = 198$ l/s about 1% mechanical losses are recorded. The gearbox efficiency changes in this area between $\eta_{getr} = 0.9 \ldots 0.99$.

10.1.5 Hydraulic Losses

The following losses were determined for the flows $Q = 0.01\ \text{m}^3/\text{s}$, $Q = 0.03\ \text{m}^3/\text{s}$, $Q = 0.06\ \text{m}^3/\text{s}$, $Q = 0.09\ \text{m}^3/\text{s}$, $Q = 0.120\ \text{m}^3/\text{s}$, $Q = 0.148\ \text{m}^3/\text{s}$,

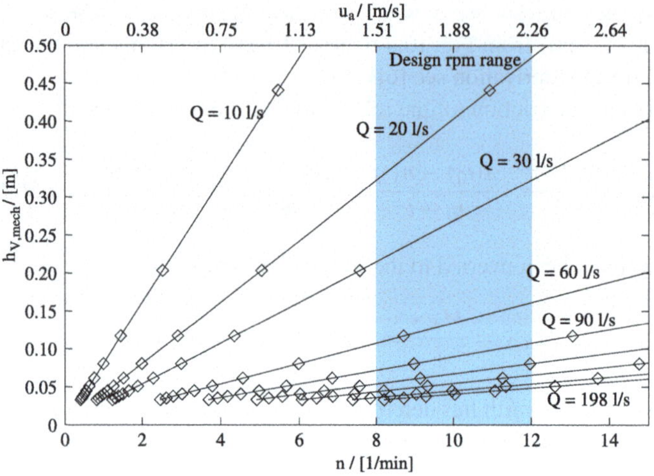

Fig. 10.5 Mechanical losses represented as loss heights. The loss height of $h_{V,mech} = 0.04$ m corresponds to a loss of 1%

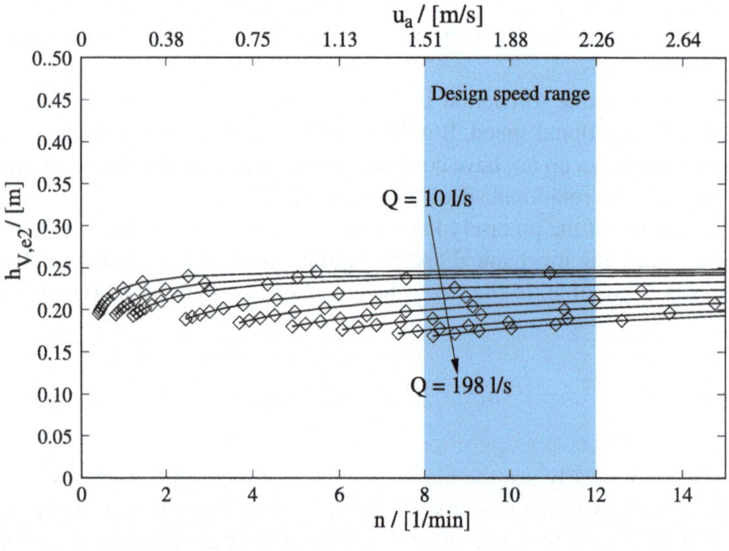

Fig. 10.6 Calculated inlet losses depending on rotational speed

$Q = 0.180$ m^3/s, $Q = 0.198$ m^3/s and $Q = 0.210$ m^3/s. In Fig. 10.6, the inlet losses of the waterwheel are shown. Here only the combined loss $h_{V,e2}$ is shown. The calculation of $h_{V,e2}$ was described in Chap. 5. The losses due to backlash on the blade have not been considered in this calculation. The backlash loss is very low in this waterwheel, as the blade curve almost corresponds to the path of the water jet. It was therefore neglected in the calculations.

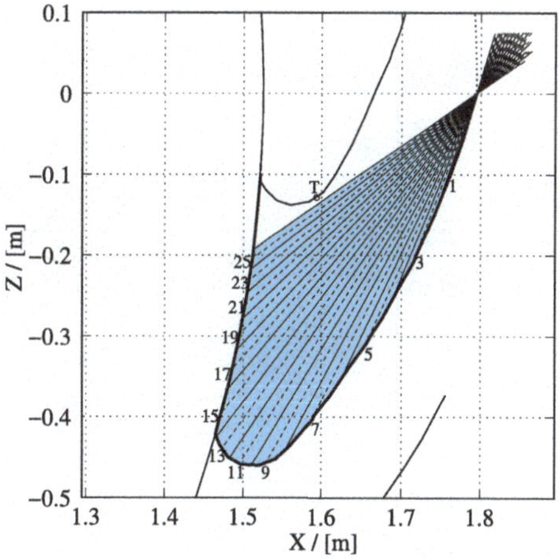

Fig. 10.7 Cross section of the cell for determining the exit losses

In Fig. 10.2 one can see that for the shown flow of $Q = 150$ l/s the water jet is completely deflected in the cell. First, the jet hits the cell bottom and is then deflected in the cell. The good bending of the cells reduces the inlet loss significantly.[4]

The inlet loss is only weakly dependent on the rotational speed. It decreases with increasing flow, as the center of gravity radius increases and thus interaction point C is moving upwards. This leads to a reduction of the distance H_C and the inlet loss decreases.

This is only true for medium flows. With very large flow rates the jet is so wide that the inlet arc is completely filled. The water is no longer deflected and this causes a shock on the blade. The cells are overfilled, the air ventilation no longer works completely and the water jet exits freely over the cells. These effects are not included in the losses at the inlet, the losses increase again for the large flow rates.

The position of the outlet opening is optimally chosen. According to Fig. 5.3, a pressure height of $h_{D2} = 0.4$ m results in a minimum loss height at $\delta_x = -0.15$ m.

The outlet losses were determined using Fig. 10.7. Since the dimensions of the cell were not exactly known, the design rules for sheet metal cells used by Müller were applied (the wheel was manufactured by the machine factory, Cannstadt. The owner of the company was W. Müller). From the drawings of Meerwarth [1], it could be inferred that the cells are well rounded. The cell was then discretized, and the exit losses were calculated using method 1 described in Chap. 5.

The method provides meaningful values up to point T (refer Fig. 10.7). A discontinuity occurs at point T. Therefore, some data points are missing, as the method is no longer applicable without errors. The calculated static and dynamic loss heights are shown in Fig. 10.8.

[4] Without considering the deflection at the inlet loss, one would have an efficiency of $\eta_{WW} = 0.75$.

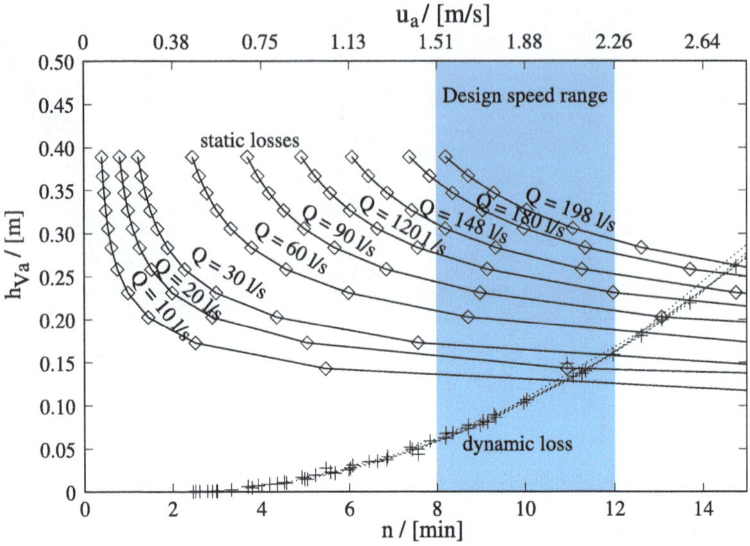

Fig. 10.8 Calculated static and dynamic exit losses

From the figures, it can be inferred that the static loss heights have a hyperbolic shape as a function of the rotational speed. Low wheel speeds only lead to heavily filled cells with high static exit loss. The dynamic losses become particularly effective when high speeds are reached. They are proportional to $0.8 \cdot u_a^2/(2 \cdot g)$. A comparison with Fig. 5.9 shows the significantly lower static losses of a curved sheet metal cell. Since curved sheet metal cells have clear advantages at the entry and at the exit of the water, the use of wooden cells should generally be avoided. In addition, the discharge capacity is lower in wooden cells as well.

10.1.6 Efficiency Curves

Figures 10.9 and 10.10 present the results of the theoretical calculation compared to Meerwarth's measurements [1]. The agreement of the calculations with the measurements is good. It is important to note that the maxima of the curves match well. For very low speeds (below $n = 4.8\ \text{min}^{-1}$), the measurements are uncertain, as the brake could not be kept in a steady state [1]. The wheel is overfilled in this case and overfilling is not included in the theoretical calculation.

The calculated high efficiencies are too large for the maximum flow. This is, as already discussed, due to the entry losses. For high speeds, the neglect of the centrifugal force effect is unfavorable in the calculations. The water is pushed towards the outer circumference and exits prematurely.

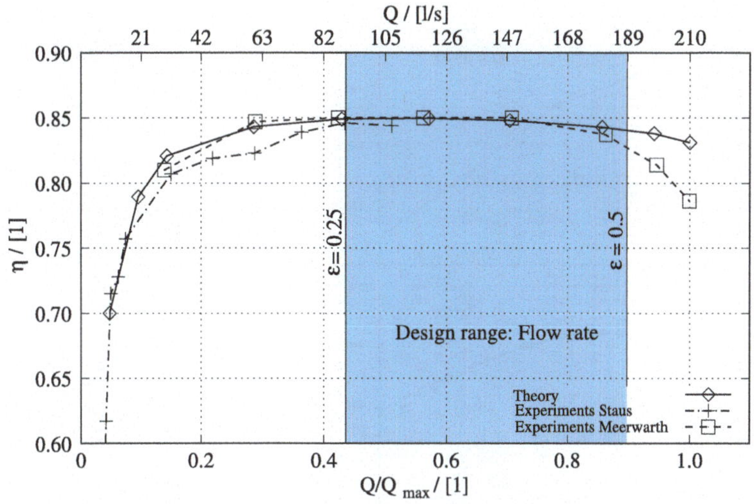

Fig. 10.9 Efficiency curve depending on the impingement according to measurements by Meerwarth [1] and Staus [3] compared with the theoretically determined efficiency curve. The waterwheel's speed is $n = 9\ \mathrm{min}^{-1}$ (efficiency with two gear stages)

Nevertheless, for the typical working range of an overshot waterwheel, meaningful values for the efficiency are calculated with the given equations. The optimal rotational speed can be determined as well. The overload area, however, is not adequately represented for the reasons mentioned. Therefore, one has to rely on empirical values, which nevertheless provide sufficient safety for the design.

In Fig. 10.10, the efficiency is plotted against the load. The data from Staus [3] and Meerwarth [1] were combined, as they cover different parts of the characteristic curve. The design flow was $Q = 150$ l/s and the maximum load was $Q = 210$ l/s. Particularly important is the result obtained for the fill factor ε. With

$$\varepsilon = \frac{60 \cdot Q}{a \cdot B \cdot 2 \cdot \pi \cdot n \cdot R_a} = \frac{60 \cdot 0.21}{0.28 \cdot 0.8 \cdot 2 \cdot \pi \cdot 9 \cdot 1.795} = 0.554$$

one obtains at $Q_{max} = 210$ l/s and the best speed of $n = 9\ \mathrm{min}^{-1}$ the maximum fill coefficient of $\varepsilon = 0.554$. The best speed corresponds to a peripheral speed at the outer circumference of $u_a = 1.69\ \mathrm{m/s}$. Up to a fill factor of $\varepsilon = 0.5$, the efficiency is uniformly high. For overshot waterwheels with rounded sheet metal cells, a fill factor of $\varepsilon = 0.5$ should be used in the design to achieve good discharge capacity. Filling factors are shown in Fig. 10.11 for different flow rates.

Exceeding this value is not recommended due to the drop in efficiency. However, falling below this value is not desirable, as the wheel width would increase at a constant speed. This increases the mass and therefore the costs for the shaft and the wheel itself.

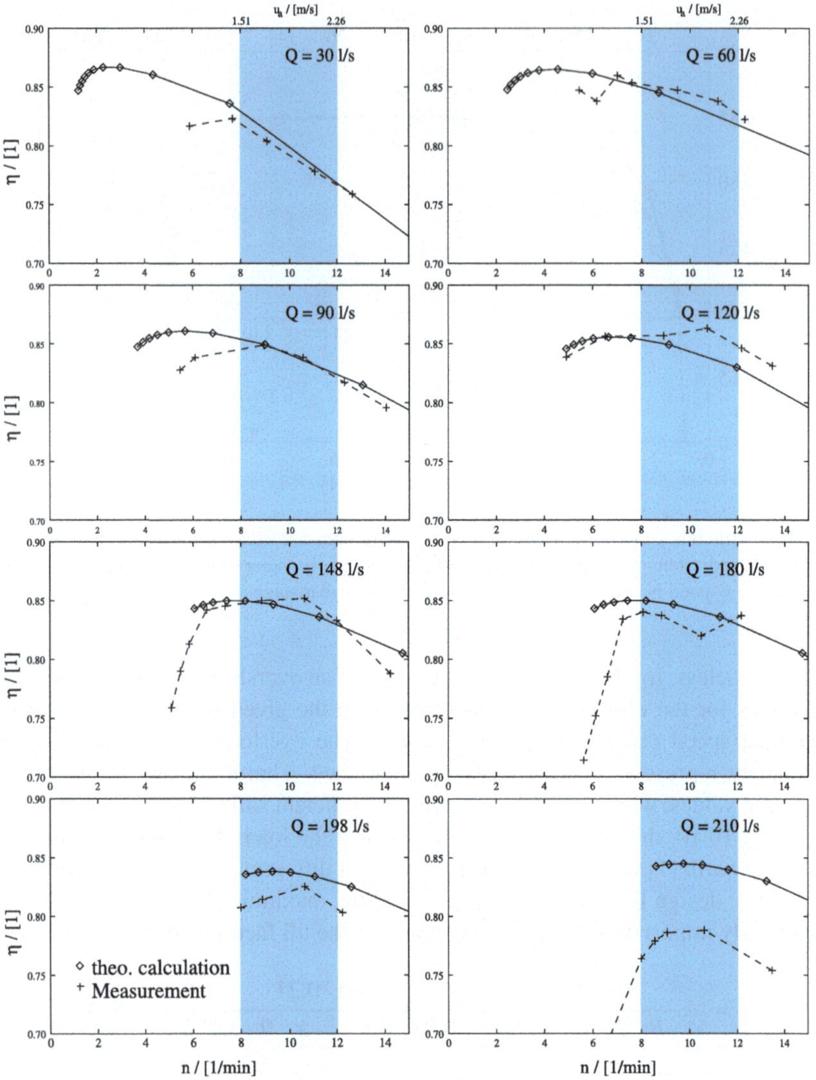

Fig. 10.10 Efficiency curves depending on the speed for different water flows $Q = 30$ l/s, $Q = 60$ l/s, $Q = 90$ l/s, $Q = 120$ l/s, $Q = 148$ l/s, $Q = 180$ l/s, $Q = 198$ l/s and $Q = 210$ l/s according to measurements by Meerwarth [1] (efficiency with two gearbox stages)

For smaller speeds than the best speed, one obtains a higher efficiency at small loads and a slightly lower efficiency at large flows.

The separation of mechanical losses was omitted, as at medium loads with a gear efficiency $\eta_{getr} = 0.99$ the efficiency only increases minimally. The mechanical losses only have a significant effect at small loads.

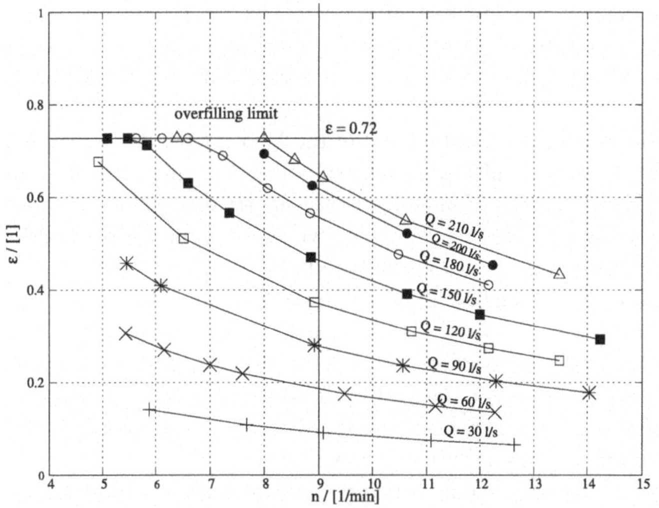

Fig. 10.11 Filling factor of the cells at different flow rates versus the rotational speed

A specific discharge capacity of $Q_s = 0.236\ \mathrm{m}^3/(\mathrm{s}\cdot\mathrm{m})$ results for this waterwheel. This value refers to $\varepsilon = 0.5$, $Q = 0.189\ \mathrm{m}^3/\mathrm{s}$ and $n = 9\ \mathrm{min}^{-1}$. The often stated specific discharge capacity of $Q_s = 0.2\ \mathrm{m}^3/(\mathrm{s}\cdot\mathrm{m})$ is somewhat too low for overshot waterwheels with sheet metal cells.

10.1.7 Summary

The test wheel of Staus and Meerwarth is a very well-constructed example of an overshot waterwheel with high efficiency.

The most important conclusions from the tests and the recalculation are:

1. After overcoming the mechanical losses (idle flow was 1.6% of the largest measured flow rate), overshot waterwheels have a nearly constant high efficiency (over 80%) at different loads. In the area of small partial loads, they are superior compared to small turbines.[5]
2. Overshot waterwheels are particularly suitable for small, strongly fluctuating flow rates at head differences of $H = 2\ldots8$ m. The design flow rate (from a flow rate duration curve) should therefore be chosen high (e.g. 50 full load days [1]).

[5] Cross-flow turbines have efficiencies of around 80% over the entire load range. Francis spiral turbines are around 90% efficient at the best point. The efficiency quickly drops below 80% in the partial load range.

3. The efficiency as a function of rotational speed is very flat. The efficiency remains constant (above $\eta = 0.8$) in the speed range of $n = 0.67 \cdot n_n \ldots 1.55 \cdot n_n$. Variable speed operation might be not very useful (additional converter losses occur).
4. The efficiency curve as a function of the load for the best speed is practically constant, as very small loads contribute almost no energy to the annual power output of the wheel. At full deflection of the jet, efficiencies of about $\eta = 0.85$ can be achieved. This type of waterwheel is particularly suitable for small hydroelectric plants with constant speed.
5. The theoretical calculation of the experimental results showed a good agreement between the measurements and the model calculations for different speeds and loads. An optimization of the efficiency is possible with the simple theoretical model. The limit ranges (overfilling at very low speeds, overload due to high flow rates, and very high rotational speed) cannot be accurately reproduced due to the simplifications used in the model.
6. The tips of the cells should touch the tailwater level for the design flow It is also possible that they are slightly submerged in the tailwater at maximum load. The drop in efficiency is compensated by the gain in head. Wading of $H_T \leq 0.1 \cdot 2 \cdot R$ should not be exceeded. It is important to accurately measure the water level in the tailwater. If there are uncertainties, one must allow a free hang ($H_T = -0.05$ m).
7. The efficiency at the best point of the overshot waterwheel remains almost constant when scaling (similarity law of Froude) the waterwheel. This was shown by the comparison of Meerwarth's [1] between the model wheel $2 \cdot R = 0.716$ m and the large wheel $2 \cdot R = 3.592$ m. Small scaled overshot waterwheels also have shown efficiencies above $\eta = 0.8$.
8. The inlet height h_{D2} and thus the inlet velocity c_{2A} have a minor influence on the efficiency. The changes in the inlet height from $h_{D2} = 0.4$ m to $h_{D2} = 0.2$ m did not lead to a drastic change in efficiency. This result also suggests good efficiencies with a water supply via an open inclined channel.
9. The passage speed was about three times the best speed $n = 9\ \text{min}^{-1}$. It is therefore significantly higher than for turbines.
10. The optimal ratio was $\frac{u_A}{c_{2A}} = 0.6$ at best speed, were c_{2A} is the horizontal component of the exit velocity and u_A is the peripheral speed of the wheel.
11. The braking torque was about twice as high as the normal torque. For waterwheels with generator coupling, this result is irrelevant, as the waterwheel-gearbox-generator arrangement is only connected to the electrical grid when the generator nominal rotational speed of the generator is reached. This situation is different when, for example, a hammer mill needs to be started from a standstill, as then the braking torque can be reached.
12. The overload limit was reached at a filling factor of $\varepsilon = 0.57$. For the design of this type of wheel, however, the slightly smaller value of $\varepsilon = 0.5$ should be used.

The conclusion is that an overshot waterwheel is very robust against changes in rotational speed and different discharge loads. By adhering to a few simple rules, quite good efficiencies can be achieved.

The waterwheel will operate for years with a constant efficiency. It is more resistant to water pollution (wastewater, sand-carrying water) compared to small turbines.

Waterwheels are well suited for self-construction. In contrast to turbines, which always require model experiments and calculations on a larger scale turbine, overshot waterwheels can be dimensioned by simple manual calculations.

In addition, waterwheels generally have little impact on the fish and other aquatic life downstream transport.

Example 10.1. ***(Planning a Hydropower Site with a Overshot Waterwheel)*** **The efficiency, power duration curve, and annual yield of a hydropower site are to be determined using the efficiency curves of Meerwarth [1]. Available are: the flow duration curve of the stream, the maximum flow $Q = 0.66\ \mathrm{m^3/s}$ according to water law, the head difference determined by experiments for different flows Q, and the generator efficiency.**

In Fig. 10.12, the flow duration curve of a stream is plotted. Since the water cannot be completely diverted from the stream, a residual water quantity of $Q = 0.075\ \mathrm{m^3/s}$ must remain in the stream. The calculations of the power output are based on the flow duration line minus the residual water. Through experiments, the head differences shown in Fig. 10.13 were determined and plotted versus the flow rate.[6] It was assumed that the upper

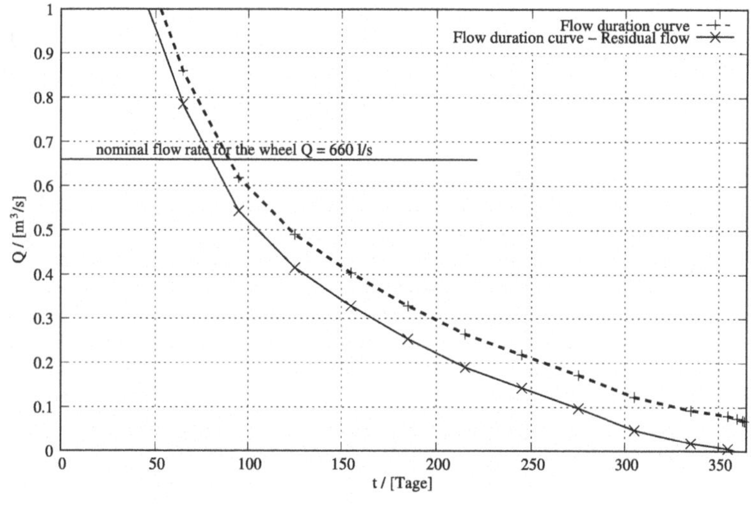

Fig. 10.12 Flow duration curve of a stream

[6] Actually, the water level in the tailrace of the channel was measured and then the water depth in the channel was determined.

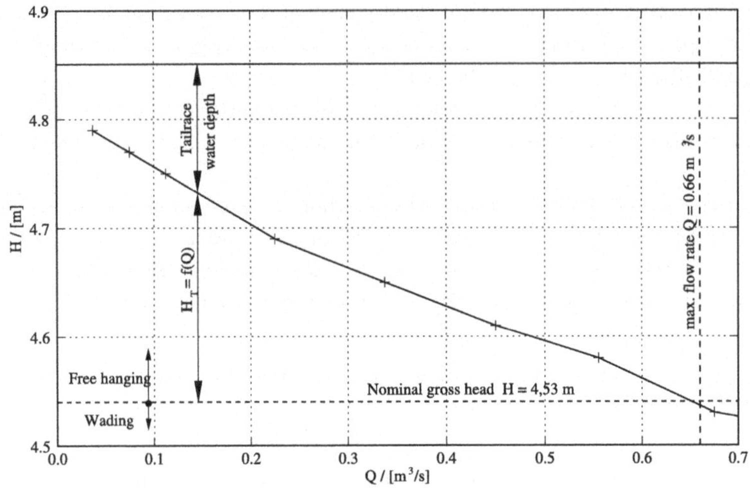

Fig. 10.13 Head difference H, immersion H_T, water depth in the channel of the lower water plotted over the flow rate

water level is constant, maintained by a sluice gate. The increase in head difference at low flow rates is caused exclusively by lowering of the tailwater level.

The head difference of $H = 4.53$ m is taken from Fig. 10.13. The overshot wheel touches the lower water level for the design flow $Q = 0.66$ m^3/s. Assuming a pressure height in front of the sluice gate of $h_{D2} = 0.4$ m and a distance of $\delta z = 0.025$, the radius of the wheel is determined:

$$R_a = \frac{H - h_{D2} - \delta z - m}{2} = \frac{4.53 - 0.4 - 0.025 - 0.005}{2} = 2.05 \text{ m}$$

The wheel should be equipped with rounded sheet metal blades like the wheel of Meerwarth [1]. Assuming a specific discharge capacity of $Q_s = 220$ l/(s · m), the wheel must have a width of $B = \frac{Q_{max}}{Q_s} = 3$ m.

This results in a conversion factor for the flow rates corresponding to the efficiency values measured by Meerwarth (wheel width $B = 0.8$ m) of $\frac{3}{0.8} = 3.75$. In Table 10.3, the converted flow rates $Q_{exp} \rightarrow Q_{site}$ are shown together with the corresponding efficiency values $\eta_{WW,exp}$ in columns 1 to 3. The experimental efficiency values still need to be converted, as no free hanging was considered in the experiments. The conversion equation is (see Table 10.3, column 6):

$$\eta_{WW} = \eta_{WW,Exp} + \frac{H_T}{H} \quad \text{if} \quad H > 4.53 \text{ m and } H_T < 0$$

$$\eta_{WW} = \eta_{WW,Exp} \quad \text{if} \quad H \leq 4.53 \text{ m}$$

Actually, one has to consider mechanical losses as well. However, this was omitted here, as only small flow rates are affected.

As soon as the wheel wades in the underwater, an efficiency is assumed that corresponds to the experimental data. Wading losses are not considered here. The power output is lower than calculated if the wheel is wading. A constant efficiency of $\eta_{gear} = 0.98^3 = 0.941$

Table 10.3 Conversion of the experimental efficiency values to the current waterwheel location

Q_{exp} [m³/s]	Q_{site} [m³/s]	$\eta_{WW,Exp}$ [1]	H [m]	H_T [m]	η_{WW} [1]
0.01	0.0375	0.700	4.79	–0.26	0.646
0.02	0.075	0.789	4.77	–0.24	0.739
0.02912	0.1125	0.820	4.75	–0.22	0.777
0.06055	0.225	0.843	4.69	–0.16	0.809
0.0888	0.3375	0.849	4.65	–0.12	0.823
0.11805	0.45	0.849	4.61	–0.08	0.832
0.1487	0.555	0.848	4.58	–0.05	0.837
0.1813	0.675	0.832	4.53	0.00	0.832
0.1989	0.7425	0.810	4.52	0.01	0.810
0.2099	0.7875	0.780	4.51	0.02	0.780

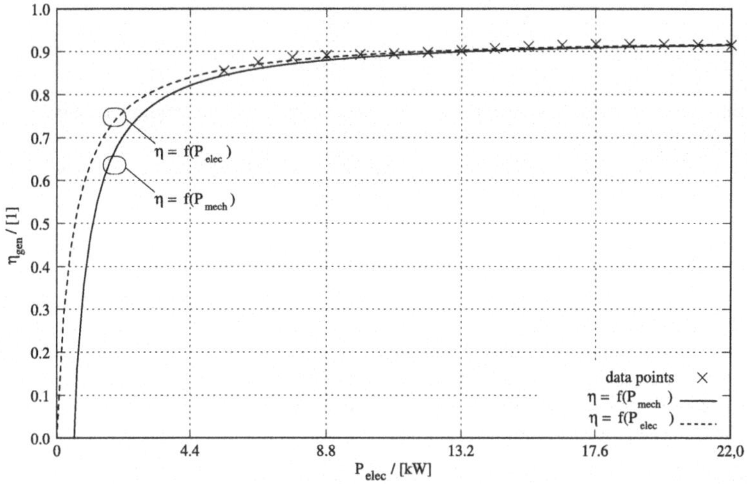

Fig. 10.14 Electrical and mechanical efficiency of the generator, plotted versus the respective power P_{elec} and P_{mech} and compared with the data points of the manufacturer

is assumed for the three-stage gearbox. Now we estimate the size of the generator. For the design head and the construction flow, we get an electrical power of the generator (η_{gen} = 0.9 estimated) of:

$$P_{elec} = 9.8 \cdot H \cdot Q \cdot \eta_{WW} \cdot \eta_{gear} \cdot \eta_{gen}$$

$$P_{elec} = 9.8 \cdot 4.53 \cdot 0.66 \cdot 0.832 \cdot 0.98^3 \cdot 0.9 = 20.65 \text{ kW}$$

We choose a generator with a rated power of 22 kW which can deliver up to 24.2 kW of electrical power with 10% overloading. The data points shown in Fig. 10.14 were known for the generator efficiency. Data points for small electrical powers are missing. These values are now determined.

The mechanical power at the generator shaft and the electrical power of the generator are linearly dependent. For this reason, the following relationships apply:

$$P_{mech} = \frac{P_{mech,max} - P_{mech,0}}{P_{elec,max}} \cdot P_{elec} + P_{mech,0}$$

$$\eta_{gen}(P_{mech}) = \frac{\frac{P_{elec,max}}{P_{mech,max}-P_{mech,0}} \cdot (P_{mech} - P_{mech,0})}{P_{mech}}$$

$$\eta_{gen}(P_{elec}) = \frac{P_{mech}}{\frac{P_{mech,max}-P_{mech,0}}{P_{elec,max}} \cdot P_{elec} + P_{mech,0}}$$

The maximum electrical generator power is $P_{elec,max} = 22$ kW. Similarly, the maximum mechanical power at the generator is known: $P_{mech,max} = \frac{22\text{ kW}}{0.915} = 24.04$ kW. With the following equation ($P_{mech,0} = 0.55$ kW) a good agreement is found between the data points and the above equation for the electrical efficiency of the generator:

$$\eta_{gen}(P_{elec}) = \frac{P_{elec}}{\frac{23.45}{22} \cdot P_{elec} + 0.55}$$

For the conversion of mechanical to electrical energy, the following equation applies:

$$\eta_{gen}(P_{mech}) = \frac{\frac{22}{23.45} \cdot (P_{mech} - 0.55)}{P_{mech}}$$

This equation will be used in the subsequent calculations. The asynchronous generator only delivers electrical power from a mechanical power of 0.55 kW, as can be seen in Fig. 10.14. At very low power levels, there is a significant drop in power, and it may happen that the generator is shut off.

Now, it is possible to determine the efficiency of the system as well as the theoretical power P_{th}, the mechanical power delivered by the waterwheel, $P_{mech,WR}$, the mechanical power present at the generator shaft, $P_{mech,gen}$, and the electrical power P_{elec}.

The following relationships are used:

$$\eta_{tot} = \eta_{WW} \cdot \eta_{gear} \cdot \eta_{gen}$$

$$P_{th} = 9.8 \cdot H \cdot Q_{site}$$

$$P_{mech,WR} = P_{th} \cdot \eta_{WW}$$

$$P_{mech,gen} = P_{th} \cdot \eta_{WW} \cdot \eta_{gear}$$

$$P_{elec} = P_{th} \cdot \eta_{WW} \cdot \eta_{gear} \cdot \eta_{gen}$$

The efficiency of the waterwheel and the generator are load-dependent, while the gear efficiency was assumed to be constant.

The values in Table 10.4 were determined with the values from Table 10.3. From a flow rate of approximately $Q = 0.05$ l/s, the generator will deliver power to the grid. The maximum system efficiency is 0.718 which is high. The performances and efficiencies from Table 10.4 are shown in Figs. 10.15 and 10.16.

To determine the power duration curve of the plant, the head and the plant efficiency are given as fit of the flow rate:

$$H(Q) = 0.18064 \cdot Q^2 - 0.5238 \cdot Q + 4.8068$$

Table 10.4 Determination of the electrical power for the hydropower site

Q_{site} [m^3/s]	H [m]	η_{WW} [1]	η_{gear} [1]	η_{gen} [1]	η_{tot} [1]	P_{th} [kW]	P_{elec} [kW]
0.0375	4.79	0.646	0.941	0.456	0.277	1.76	0.49
0.075	4.77	0.739	0.941	0.726	0.505	3.51	1.77
0.1125	4.75	0.777	0.941	0.803	0.585	5.24	3.06
0.225	4.69	0.809	0.941	0.873	0.664	10.34	6.87
0.3375	4.65	0.823	0.941	0.895	0.693	15.38	10.66
0.45	4.61	0.832	0.941	0.906	0.709	20.33	14.41
0.555	4.58	0.837	0.941	0.912	0.718	24.91	17.90
0.675	4.53	0.832	0.941	0.916	0.717	29.97	21.50
0.7425	4.52	0.810	0.941	0.918	0.701	32.89	23.07
0.7875	4.51	0.780	0.941	0.918	0.678	34.81	23.59

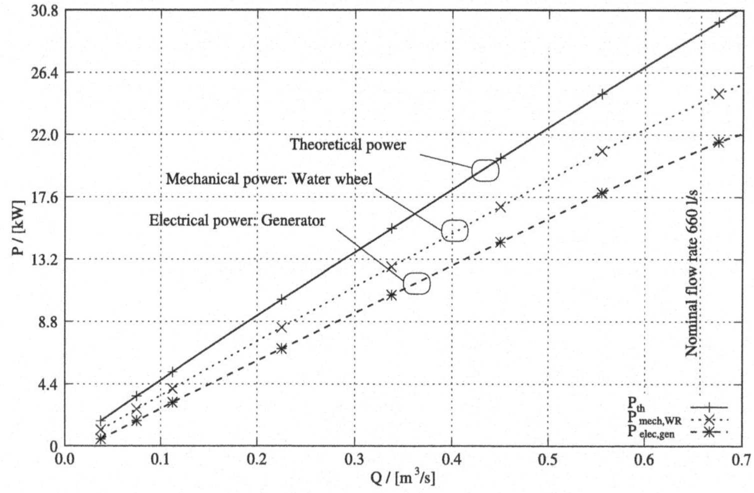

Fig. 10.15 Theoretical, electrical and mechanical power, plotted versus the flow rate

$$\eta_{tot}(Q) = a \cdot \exp(-b \cdot Q) + c \cdot Q^2 + d \cdot Q + e \quad \text{with}$$
$$a = -0.93441 \quad b = 29.0231 \quad c = -0.41017$$
$$d = 0.487838 \quad e = 0.5744$$

The discharge duration curve is converted into a power duration curve with these equations above. First, the theoretical power $P_{th} = 9.8 \cdot H \cdot Q$ is determined for each data point and then the theoretical power is multiplied by the plant efficiency.

The result of these calculations is shown in Fig. 10.17. The hydropower plant uses the design flow rate for a period of 75 days.

Below 1 kW of electrical power, the plant is shut down. From the graph, it can be seen that the waterwheel is idle for 60 days in an average year due to low water levels. An adaptive residual water flow could improve this situation.

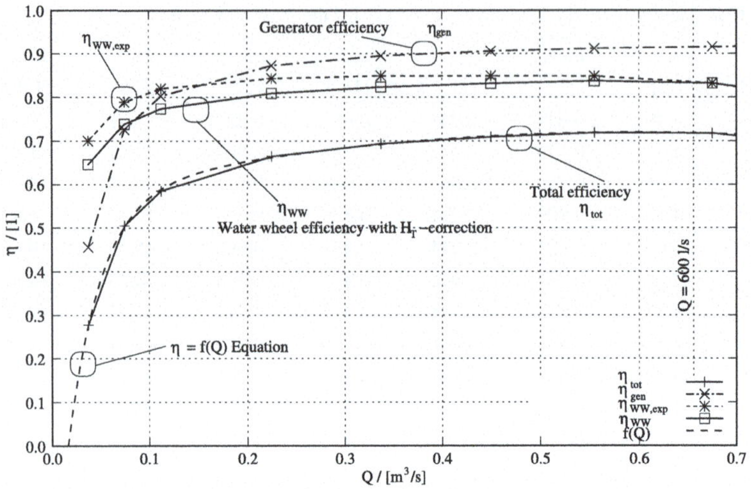

Fig. 10.16 Efficiency of the hydropower plant, plotted versus the flow rate

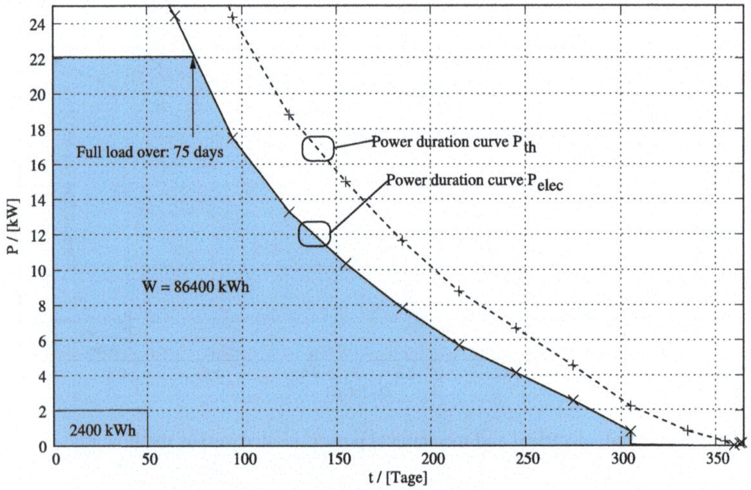

Fig. 10.17 Power duration curve and average annual energy of the hydropwoer plant

By counting the squares, the average annual energy of the plant is obtained. Each rectangle bordered by dotted lines corresponds to 2 kW · 50 · 24 h = 2400 kWh. The incomplete rectangles are accordingly supplemented. This results in an average annual energy of 36 · 2400 kWh = 86,400 kWh. The estimate of the annual energy at the design flow of 200 full load days ($W = 200 \cdot 24\ \text{h} \cdot 22\,\text{kW} = 105{,}600$ kWh) is too high for this location.

Fig. 10.18 Front and side view of the overshot waterwheel in the hydraulic laboratory of the University of Wisconsin [8]. The figure is a low quality copy from Weidner's paper

10.2 Tests on the Overshot Waterwheel at the University of Wisconsin

In 1913, various test series were conducted on an overshot waterwheel at the University of Wisconsin by Weidner [7]. The experimental setup will be briefly described. The measurement arrangement is shown in Fig. 10.18. The water was supplied to the wheel via a 16-inch pipe (0.406 m). Two steam-driven centrifugal pumps were used, which had a maximum flow rate of $Q = 0.198\ \text{m}^3/\text{s}$. The water was drawn directly from Lake "Mendota". The 16-inch pipe ended directly in an open steel tank, which served as a balancing tank. The steel tank had the following dimensions: $H = 0.914$ m, $B = 0.756$ and $L = 1.067$ m. The water level in the tank could be measured via a glass tube with a scale.

The outlet from the steel tank was controlled via a sluice gate. The opening width of the gate sluice was $b_0 = 0.610$ m and thus slightly narrower than the tank itself. The maximum jet width was $s_0 = 0.146$ m. An open, straight outflow channel followed the sluice gate opening. This had the same width as the sluice of 0.610 m and a length of 0.711 m. This channel ended $\delta x = -0.21$ m before the upper wheel vertex.

The waterwheel was supplied by the Fitz-Waterwheel Company, Hanover, Pennsylvania, as a hybrid steel construction (cells, wheel rims riveted from steel and shaft, rosettes made of cast iron) and had a diameter of $D_a = 3.048$ m (10 feet). The number of cells was $z = 32$ and the wheel had a rim width of $a = 0.2$ m.

The capacity of a fully loaded cell was $V_z = 0.038\ \text{m}^3$. This volume, when applying the optimal speed according to Eq. (8.3) for this wheel: $n_{opt} = 11.75/\sqrt{R_a} = 11.75/\sqrt{1.524\ \text{m}} = 9.52\ \text{min}^{-1}$ results in a flow rate of: $Q = \varepsilon \cdot V_z \cdot z \cdot n/60 = 0.5 \cdot 0.03806 \cdot 32 \cdot 9.52/60 = 0.097\ \text{m}^3/\text{s}$, assuming a filling factor of $\varepsilon = 0.5$. The specific discharge capacity is thus $Q_s = Q/B = 0.128\ \text{m}^3/(\text{s} \cdot \text{m})$

and is therefore relatively low, because the rim width with $a = 0.2$ m was chosen to be relatively small in order to achieve good efficiencies.

Two holes were added at the bottom of the cells to ensure proper ventilation. These had a diameter of 1.587 cm and were located at the outer ends of the cells. They served to empty the cells when stationary, to prevent the suction of water when wading in the underwater and to improve the ventilation of the cells at the inlet.

A segmented spur gear was mounted on the outer rim of the wheel. This had 168 teeth and engaged with a smaller gear with 42 teeth. The transmission ratio is thus $i = 168/42 = 4$. The waterwheel was installed on a concrete foundation and hung freely over a 1.216 m wide tailwater channel. Approximately 15 m downstream, a standard rectangular weir without lateral contraction of the jet (weir width $b = 1.069$ m) was installed. The waterwheel's discharge capacity could thus be determined by measuring the overfall height of the weir.

The power measurement was carried out using two Prony brake systems. The brake drum of the first brake was mounted directly on the wheel's shaft.

However, during measurements, it was observed that at low rotational speeds ($n < 12$ 1/min), a stable braking state could not be achieved, as the brake caused the brake drum and wheel to come to an abrupt halt. Therefore, the first brake was not used in the remaining tests and was dismantled. The second brake was mounted on the shaft of the gear wheel (with 42 teeth) and allowed stable measurements down to very low rotational speeds. The following results were obtained in a large number of measurement series [8] which are not shown here in detail:

- the maximum efficiency (at low speed) was 89%.
- the transmission loss (including losses in the bearings) was 3%. This increased with speed and decreased at high power at the wheel shaft.
- the efficiency of the wheel increased when the inlet speed was reduced. This only applies as long as the inlet speed is higher than the peripheral speed of the wheel.
- a variation of the flow only has a minor influence on the efficiency, as long as the fill factor $\varepsilon = 0.5$ is not exceeded.
- to achieve maximum efficiency, the ratio of $u_a/c_{2A} = 0.9$ should be chosen and the point of water entry into the waterwheel should be placed as close as possible to the upper vertex of the wheel. The entry point depends on the distance of the outlet opening to the wheel apex and must be adjusted for each waterwheel.
- a loss of efficiency of about 4% is observed (at a optimum rotational speed of $n_{opt} = 9.52$ min^{-1}) if the rim is submerged in water and the tips of the sheet metal cells just touch the water surface.

Only the wading experiments are of interest here, as they complement the measurements of Staus [3] and Meerwarth [1].

The results of the wading experiments are shown in Fig. 10.19. A considerable loss of efficiency occurred. For example, at a wading depth of $H_T = 0.1 \cdot 2 \cdot R_a$, the efficiency at $n_{opt} = 9.52$ min^{-1} is reduced to about 36%. This is due to the fact that the wheel was tested at a flow rate of only $Q = 63$ l/s. At a real location, wading will only occur with proper design when the maximum flow rate (for this wheel this corresponds to $Q = 0.097$ m^3/s) is exceeded.

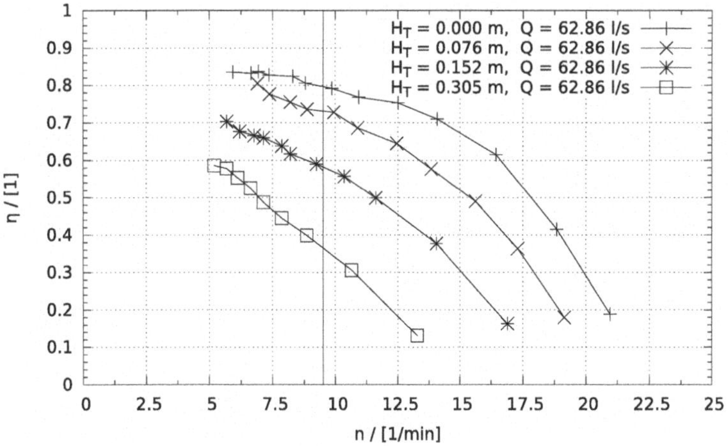

Fig. 10.19 Efficiency of the wading experiments for different immersion depths $H_T = 0.0$ m, $H_T = 0.076\,\text{m} = 0.025 \cdot 2 \cdot R_a$, $H_T = 0.152\,\text{m} = 0.05 \cdot 2 \cdot R_a$ and $H_T = 0.305\,\text{m} = 0.1 \cdot 2 \cdot R_a$

The hydraulic wattage loss remains the same, but the theoretically available hydraulic power and thus the efficiency of the wheel is then higher. This can be converted using the relationship $\eta_{WR,Wading} = \eta_{WR} - \frac{P_{V,Wading}}{P_{th}}$ by separating the constant wattage loss $P_{V,Wading}$ and converting it to other flows. According to Fig. 10.19, the wading of the wheel can thus be used to limit the power of overshot waterwheels.

10.3 Efficiency Measurement at a Hydropower Site

10.3.1 Description of the Hydropower Site Stedten

The waterwheel in Stedten, Thuringia is located on a diversion channel of the river Ilm. The water collection weir was renovated in 2005, and a new waterwheel was installed. The overshot waterwheel is schematically represented in Fig. 10.20, and the geodetic elevations can be found in Table 10.5. A photograph is shown in Fig. 10.35 All essential height specifications are also indicated in the figure. All other information about the waterwheel, such as the peripheral speed, can be found in Table 10.7.

The monthly mean flow rates of the river Ilm are given in Table 10.6 (Source: Thuringian State Institute for Environment and Geology). The MQ and MNQ mean flow rate values very rarely fall below the construction flow of 0.44 m^3/s. The plant could always operate at full load or even handle a higher flow. The advantages for a power measurement at this plant were:

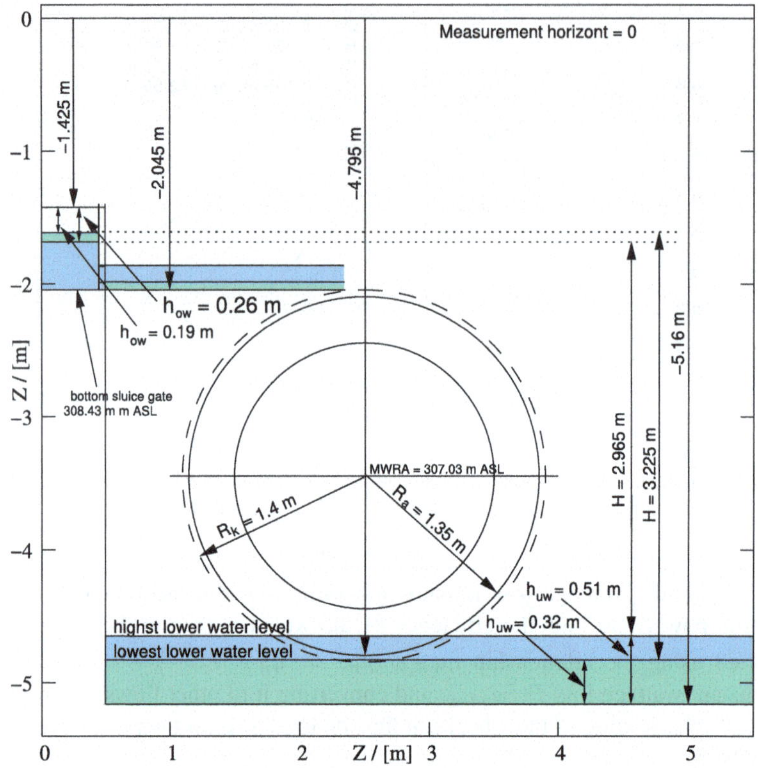

Fig. 10.20 Schematic representation of the measured heights at the waterwheel location of the plant in Stedten. The data points of the measurement series 2 for the rounded sluice gate (measurement point 1: Opening width of the sluices $s_0 = 0.055$ m and measurement point 7: Opening width of the sluices $s_0 = 0.175$ m) from Table 10.11 are shown

1. All details about the plant were known (installation drawings of the hydropower site were available from the year 2005 (including drawings for the weir shown in Fig. 10.33 and the inlet sluice gates shown in Fig. 10.34 etc.).
2. The flow rate of $Q = 1.8$ m^3/s was very constant over the measurement period of seven hours. It was below the annual mean discharge of $Q_{MQ} = 2.45$ m^3/s. This was checked using the water gauge level in Gräfinau-Angstedt (source: website of the Thuringian State Institute for Environment and Geology).
3. The water flow could be determined using three methods. On the one hand, the flow was determined using a submerged sluice gate with a sharp-edged, well-rounded leading edge. For testing purposes, the sharp-edged sluice gate was converted into a sluice gate with a well-rounded leading edge by installing a pipe. The rubber lip of the sharp-edged sluice led to a good seal of the mounted steel pipe against the sharp-edged sluice exit edge. On the other hand, a wooden broad-crested weir with a inclined chute was used. The broad-crested weir was

Table 10.5 Geodetic information about the waterwheel location in Stedten

Parameter	Value	Remark
GPS Longitude	11° 10.94E	
GPS Latitude	50° 50.908N	
GPS Altitude Measurement	306 m.a.s.l.	Accuracy $\pm 2m$
Design flow rate	$Q = 0.44$ m^3/s	Currently up to $Q = 0.63$ m^3/s
Geodetic head	$H = 3.12$ m	Currently $H_{Geo} = 2.95 \ldots 3.225$ m
Electrical power	$P_{elec} = 8.75$ kW	Assumed efficiency 0.65, currently up to 11.0 kW
Geodetic elevation of the bottom of the Sluice Gates	308.43 mASL	Reference point of the measurement
Geodetic elevation of the Waterwheel shaft midpoint	307.03 mASL	MWRW
Lower wheel vertex	305.68 mASL	Lowest point of the blade edges
Upper water level	308.75 mASL	Indicated as OWS
Lower water level	305.63 mASL	Indicated as UWS
Channel bed in the tailwater	305.28 mASL	
Weir crest level	308.83 mASL	

dimensioned with the software WinFlume. For this, a wedge-shaped arrangement was installed behind the fully open sluice gate.

4. The condition of the waterwheel was excellent, as the wheel had only been in operation for a year. The waterwheel had angled sheet metal blades similar to the waterwheel presented in the calculation examples (Groß Werzin) but a slightly smaller wheel diameter compared to Meerwarth [1].

At this point, special thanks are extended to Mr. U. Müller (the owner of the hydropower site) and Dr.-Ing. Treiber (Thüringer Wasserkraftanlagen GmbH) who built the overshot waterwheel in 2005.

10.3.2 The Measurement

The measurement was conducted on March 24, 2006 and started at 10:00 AM. The participants were the owner of the plant U. Müller, Dr.-Ing. G. Treiber, and the author.

At the beginning of the measurement, the plant was operated with the sluice gate fully pulled out of the water. A sufficiently high pressure head h_{D2} could not build up in front the sluice gate. The sluice of the inlet structure at the weir was lowered to limit the water flow. This condition led to a power output of $P_{elec} = 5$ kW displayed by the wattmeter in the power house.

Table 10.6 Main flow rates [m^3/s] 1923–2002 above the waterwheel location in Stedten: Name of the gauge: Gräfinau-Angstedt, Station number: 57289, Location above the confluence: 108.0 km, precipitation area: 154.8 km^2, Gauge zero point: 407.530 m.a.s.l.

Month	NQ	MNQ	MQ	MHQ	HQ
November	0.220 (1991)	1.07	2.47	7.31	49.2 (1940)
December	0.180 (1953)	1.21	3.19	10.2	47.7 (1947)
January	0.230 (1954)	1.31	3.40	10.4	55.6 (2002)
February	0.210 (1963)	1.45	3.34	9.37	69.3 (1946)
March	0.210 (1963)	1.60	3.79	10.7	60.8 (1981)
April	0.540 (1960)	2.09	4.20	9.51	49.3 (1994)
Winter	0.180 (1953)	0.664	3.40	21.0	69.3 (1946)
May	0.280 (1943)	1.09	2.13	5.27	18.0 (1969)
June	0.140 (1954)	0.776	1.68	5.32	23.2 (1972)
July	0.190 (1934)	0.653	1.31	4.38	14.7 (1996)
August	0.180 (1947)	0.550	1.06	4.94	79.6 (1981)
September	0.160 (1928)	0.602	1.20	3.60	25.7 (1998)
October	0.220 (1933)	0.705	1.72	4.96	24.8 (1960)
Summer	0.140 (1954)	0.423	1.52	10.4	79.6 (1981)
Year	0.140 (1954)	0.382	2.45	22.0	79.6 (1981)

This operating condition led to strong lateral splashing of the wheel, as the exit speed from the channel was too low compared to the speed of the overshot wheel. After opening the sluice gate at the inlet structure, the lateral splashing became much stronger, as now more water could flow to the wheel. Since the wheel is located very close to the machine house, the splashing caused the wall to become soaked, and the water seeped into the gearbox area, forming a puddle at its base.

The lateral water exit stopped only when the sluice was lowered enough to create a sufficient pressure head of $h_{D2} = 0.37$ m, with a sluice opening width of $s_0 = 0.175$ m. The power increased through this measure to $P_{elec} = 10$ kW. The measurements started after a proper operating state for the hydropower plant was established.

The work program consisted of the following steps:

1. Inspection of the site.
2. Installation of the measuring equipment.
3. Surveying work, including the attachment and calibration of the staff gauges in both the upper and lower waters, was carried out during the plant's operation. The bottom of the sluice was chosen as the reference point.
4. Measurement of the head difference and the pressure height during operation.
5. Change of the inflow conditions by increasing the pressure height h_{D2}.
6. Repeated recording of the pressure height, the opening width of the sluice, determination of the contraction factor $\psi = h/s_0$, reading of the staff gauges, measurement of the electrical power by speed measurement at the generator, reading of the power meter in the switch cabinet as well as current and voltage with the help of an oscilloscope.

Table 10.7 Geometric dimensions of the overshot waterwheel at the hydropower site Stedten

Parameter	Value	Unit
Geodetic head difference H	3.225 ... 2.965	[m]
Flow rate	0.15 ... 0.63	[m^3/s]
Center of waterwheel shaft M	307.03	[m.a.s.l.]
Outer radius R_a	1.35	[m]
Number of blades z	28	[1]
Filling factor ε	0.168 ... 0.514	[1]
Nominal flow rate Q_{nom}	0.44	[m^3/s]
Pressure head h_{D2}	0.435 ... 0.36	[m]
Width of the sluice gate b_0	2.0	[m]
Wheel width B	2.1	[m]
Rim width a	0.35	[m]
Submersion H_T	−0.05	[m]
Specific discharge capacity Q_s	0.220	[m^3/(s · m)]
Rotational speed n	10.88	[min^{-1}]
Peripheral speed u_a	1.538	[m/s]

7. Installation of a steel pipe to obtain a well-rounded inflow edge of the sluice gate.
8. Recording of the same parameters as described in work point 6.
9. Installation of an arrangement that simulates an inflow over a broad-crested weir with subsequent chute (flume designed with Winflumes).
10. Opening of the sluice gate to that extend that a free-flow was established towards the overshot waterwheel.
11. The measurement were repeated as described in point 6. The pressure height and the overflow height of the broad-crested weir was determined instead.
12. Determination of the waterwheel's overspeed.

The weather was sunny and calm. The average flow rate $Q_{MQ} = 1.8$ m^3/s in the Ilm was slightly below mean flow rate of $Q_{MQ} = 2.45$ m^3/s. After setting up the leveling instrument, the power measurement instruments were installed. Furthermore, the type plate of the generator was copied (Table 10.7).

In Tables 10.8 and 10.9 one can find the information about the gear and the three-phase asynchronous generator. In Fig. 10.21 the characteristic curves of the generator are shown. It is important to mention that there is an almost linear relationship between the mechanical and the electrical power. This is true as well as for the slip $s = 1 - n_{syn}$ and the electrical power of an asynchronous generator.

By measuring the rotational speed at the generator, the slip can be determined. With the slip, the mechanical input power P_{mech} can be determined from the characteristic curves in Fig. 10.21. and the electrical output power P_{elec} can be extracted. The generator only delivers electrical output power from approximately 1 kW input power on. Then a staff gauge was calibrated in the headwater, the center of the waterwheel shaft was measured, and the height of the tailwater water level

Table 10.8 Data of the three-stage gear and rotation speeds in the drive system under load at an electrical power of $P_{elec} = 10$ kW (see measurement series 1, line 1, Table 10.10)

	Diameter [1]/[m]	Diameter [1]/[m]	Transmission [1]	Rotation speed [min^{-1}]
Waterwheel speed				10.88 min^{-1}
2-Stage gear			1 : 24.105	10.88 : 262.26 min^{-1}
Belt drive	$d_1 = 0.630$ m	$d_2 = 0.160$ m	1 : 3.9375	262.26 : 1030 min^{-1}
Generator speed				1032.6 min^{-1}

Table 10.9 Nameplate of the generator and manufacturer's specifications

VEM Electric Motors Factory Wernigerode		
Δ / Υ 400/690 V	35.5/20.5 A	
26.5 Ps/19.5 kW	$\cos\varphi = 0.79$	$\eta_{gen} = 0.87$ (datasheet)
$n_{Gen} = 1032$ min^{-1}	$f = 50$ Hz	
IP55	$m = 0.175$ t	
$S = 24.8$ kVA		

was determined. The measurements have an accuracy of ±0.005 m. This is due to the fluctuations of the surface water caused by waves (especially in the tailwater). Furthermore, the concrete channel bed of the sluice gate and the end of the inlet channel just before the waterwheel were measured. Another reference point was the concrete side wall of the channel in the headwater. The inlet channel has no gradient and was flush with the side crowns of the waterwheel.

The diversion channel was also measured from the inlet structure at the weir to the entrance of waterwheel. The slope of the water level for the whole channel has low gradient with 0.1 m over a length of 275 m. This was noticeable in that the pressure head in front of the sluice significantly decreased with larger flows. However, the normal storage level for Q_{MQ} in the river Ilm with closed sluice gate was reached.

Then, the sluice gate at the overshot waterwheel was opened to $s_0 = 0.175$ m, and the following measurements were taken in the same manner each time:

- Reading the gauge level in the headwater h_{ow},
- Reading the gauge level in the tailwater h_{uw},
- Determining the opening width of the sluice gate plate s_0,
- Re-measuring the pressure head as a value to be cross checked h_{D2},
- Measuring the speed at the generator (rotational frequency f) and
- Reading the electrical power at the wattmeter P_{elec}.
- Furthermore, the voltage (with a measuring transformer), the current (with a current clamp) and the power factor were determined with an oscilloscope. However, since the current measurements were very inaccurate, these values were discarded for further use.

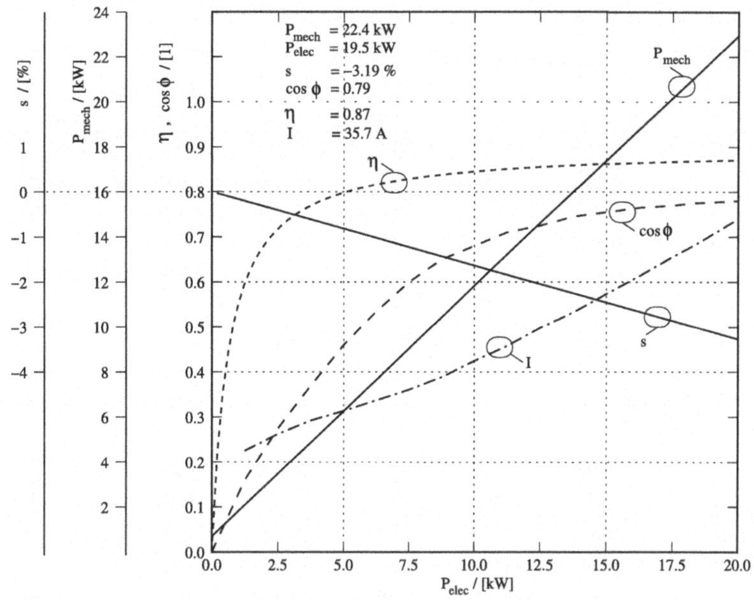

Fig. 10.21 Slip s, power factor $\cos\varphi$, efficiency η and mechanical power P_{mech} of the tested generator, plotted over the electrical power (delta connection)

- The planned measurement with the grid meter also had to be omitted, as the plant is operated in parallel with the grid and not all consumers were switched off.

Unfortunately, the value of the jet contraction ($\psi = h/s_0$) was not determined for each operating measurement point.[7] The contraction coefficient was only determined at the beginning and end of the measurement. For the sharp-edged sluice gate, the measured value was $\psi = 0.65$. This is quite close to the coefficient of $\psi = 0.62$ given by Bollrich [9]. However, the determined efficiencies are sensitive to changes in the value of ψ and it would have been better to determine ψ for each individual operating point.

The sluice gate was closed step by step then beginning with the maximum opening The step size was $s_0 = 0.02$ m. After reaching a steady state in the approach channel the measurements described above were repeated three times and averaged.

Then the sluice gate plate was re-arranged to achieve a good rounding of the leading edge. For this purpose, a pipe steel with welded flat steels rods was mounted on the sluice gate plate with clamps. The rubber lip of the sluice gate plate sealed the space between the steel pipe and the lowermost plank of the sluice gate. There was no additional leakage for this reason. After this conversion, the now rounded

[7] Here, the symbol h refers to the water depth of the jet flowing toward the waterwheel. The value was measured one meter away from the sluice gate exit.

Table 10.10 Measurement series 1: Raw data of the power measurements with sharp-edged sluice gate plate

No	h_{OW} [m]	h_{UW} [m]	s_0 [m]	h_{D2} [m]	f [1/s]	P_{el} [kW]	Remark
1	0.25	0.48	0.175	0.37	17.21	10.0	$\psi = 0.65$
2	0.24	0.46	0.155	0.38	17.165	9.0	
3	0.23	0.44	0.135	0.385	17.125	8.5	
4	0.22	0.42	0.115	0.40	17.06	7.8	
5	0.21	0.39	0.095	0.415	17.01	6.5	
6	0.20	0.35	0.075	0.43	16.93	5.5	
7	0.19	0.32	0.055	0.44	16.85	3.6	$\psi = 0.63$

Table 10.11 Measurement series 2: Raw data of the power measurements with well-rounded sluice gate plate

No	h_{OW} [m]	h_{UW} [m]	s_0 [m]	h_{D2} [m]	f [1/s]	P_{el} [kW]	Remark
1	0.26	0.51	0.175	0.36	17.22	11.0	$\psi = 0.8$
2	0.25	0.48	0.155	0.37	17.19	10.5	
3	0.245	0.45	0.135	0.38	17.168	10.0	
4	0.225	0.43	0.115	0.395	17.12	9.0	
5	0.215	0.40	0.095	0.41	17.06	8.0	
6	0.20	0.36	0.075	0.435	17.00	6.8	
7	0.19	0.32	0.055	0.435	16.90	5.0	$\psi = 0.94$

sluice was again opened between $s_0 = 0.055 \ldots 0.175$ m, and the measurements were carried out as described above.

Even during the measurement, it was noticed that the installation significantly improved the inflow behavior of the overshot waterwheel. The jet was less contracted compared to the sharp-edged arrangement (the contraction coefficient was $\psi = 0.92$ at its best). It was possible to direct significantly more water at the same opening width of the sluice into the waterwheel.

The power output increased for the maximum opening width $s_0 = 0.175$ m to 11 kW (compared to 10 kW with a sharp-edged leading edge). In addition, the tendency for strong lateral spraying decreased. Nevertheless, a pulsating escape of the compressed air in the cell space was noticeable at high flow rates. This led to a loud, pounding noise and announced the beginning overload of the waterwheel. In this case the air could no longer be sufficiently ventilated from the cell chambers.

Each measurement series lasted about five to ten minutes. This period was sufficient to reach a steady value for the lower and upper water level. The results of the first two measurement series are summarized in Tables 10.10 and 10.11.

During the measurements with the sluice gate, it was noticed that the pressure height in front of the sluice h_{D2} decreased with a larger opening width of the sluice and thus a larger flow. The reason for this is the low gradient of the diversion

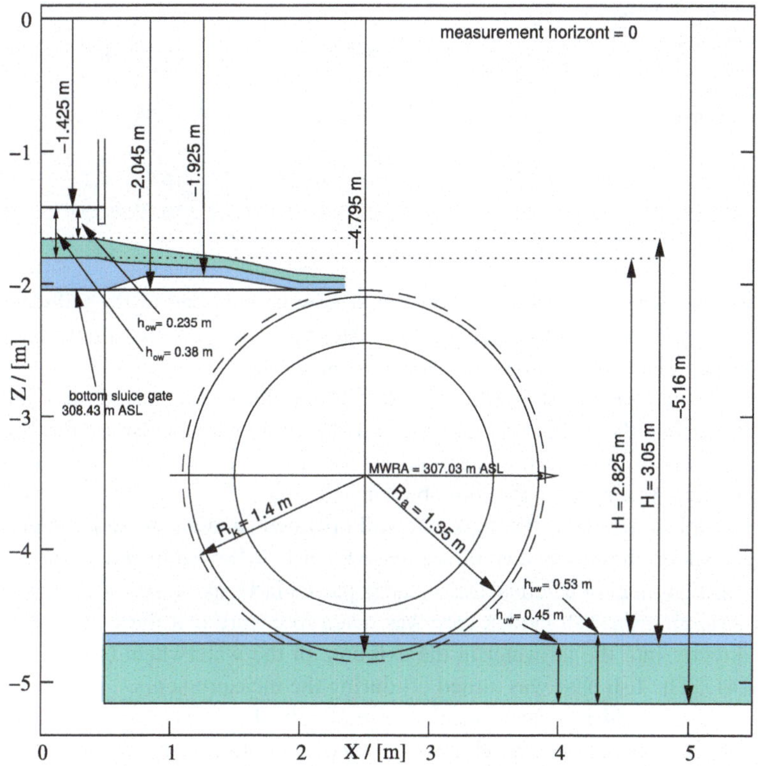

Fig. 10.22 Schematic representation of the measured heights for the waterwheel of the plant in Stedten (variant with inflow over a broad-crested weir). The data points of measurement series 3 (Measurement point 1: Overflow height of the sluice $h_{ov} = 0.265$ m and Measurement point 5: Overflow height of the sluice $h_{ov} = 0.120$ m) from Table 10.12 are shown

channel in the headwater. To achieve a certain flow rate, a corresponding gradient of the water surface line must be established. This was the cause for the decrease in pressure height h_{D2} in front of the sluice at higher flow rates. However, the planned head difference was achieved for the nominal flow rate ($Q_{nom} = 0.44\ \text{m}^3/\text{s}$).

A measuring channel was chosen as a third experimental setup, which corresponds to a broad-crested weir with inclined chute. The arrangement was installed in the inflow channel after the fully opened sluice gate and fixed with wooden wedges. It is schematically shown in Fig. 10.22. This arrangement provided the best inflow behavior. There was no contraction of the jet at the guides of the sluice and the water stream flowed as smooth jet into the waterwheel. The pounding noise was absent in this arrangement.

To vary the flow over the overflow height, it was necessary to open the free wheel sluice gate directly beside the overshot water. The sluices of the inlet structure of the diversion channel could not be used for this purpose.

Table 10.12 Measurement series 3: Raw data of power measurements with broad-crested weir and subsequent chute

Nr	h_{OW} [m]	h_{UW} [m]	f [1/s]	P_{el} [kW]	Remark
1	0.235	0.45	17.15	9.5	
2	0.265	0.48	17.08	8.0	
3	0.28	0.50	17.03	7.2	
4	0.315	0.52	16.92	5.2	
5	0.38	0.53	16.79	2.5	

However, since a larger water flow could now be supplied to the tailwater, the tailwater level rose sharply, causing the waterwheel to become submerged. The measurement data values are summarized in Table 10.10.

Finally, the overspeed reached by the waterwheel was tested. It could be determined that the overspeed depends on the impingement. Since the plant is equipped with an overspeed protection, the solenoids on the sluice were triggered before a stable overspeed state could be reached.

In parallel to the plant, fish tanks are still operated[8] and a fish ladder is in place. This fish ladder consists of a spiral staircase for fish, followed by a pool pass.

The arrangement of a helical ascent aid is particularly space-saving as a fish pass. When arranging the fish ladder, care was taken to ensure that the guide jet directs water directly into the tailwater in the exit area of the waterwheel (see Figs. 10.40 and 10.41). The fish pass was turned off during the measurements.

Since a fish ladder is also installed at the weir, the plant can be described as exemplary in terms of ecological fish friendly construction.

10.3.3 Flow Calculation for the Sluice Gate Measurements

First, the flow is calculated using the following equation:

$$Q = b_0 \cdot \psi \cdot s_0 \cdot c_2 = \frac{\psi}{\sqrt{1 + \frac{\psi \cdot s_0}{h_{D2}}}} \cdot b_0 \cdot s_0 \cdot \sqrt{2 \cdot g \cdot h_{D2}}$$

The calculation results with the input values taken from Tables 10.10 and 10.11 are shown in Fig. 10.23. The contraction coefficient $\psi = 0.65$ was used for the calculations. This value was measured for the sluice gate plate with sharp-edged inflow edge. The contraction coefficient was linearly interpolated for the rounded sluices between the two known values at the beginning and end of the measurement series.

[8] The plant operator uses a part of the generated power for the operation of a fish farming plant and thus optimally uses the power of his plant in grid parallel operation.

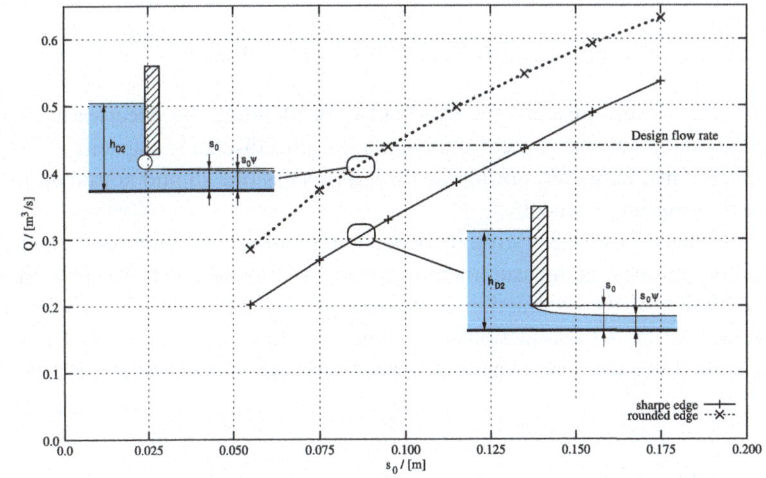

Fig. 10.23 Flow, plotted over the opening width s_0 of the sluice gate

The equation for this interpolation is:

$$\psi = -1.1667 \cdot s_0 + 1.004$$

Both variants were able to supply the design flow rate of $Q = 0.44\ \mathrm{m}^3/\mathrm{s}$ to the waterwheel. However, the respective opening widths to discharge the design flow are very different for the two sluice gate variants.

For the steel pipe variant the jet width is narrower, and the ventilation of the cells of the waterwheel improves significantly (less noise).

E.g. for measurement series 2 measurement no. 1 the following results are calculated:

$$b_0 = 2\ \mathrm{m} \qquad h_{D2} = 0.36\ \mathrm{m} \qquad s_0 = 0.175\ \mathrm{m}$$

$$\psi = -1.166 \cdot 0.175 + 1.004 = 0.8$$

$$Q = \frac{\psi}{\sqrt{1 + \frac{\psi \cdot s_0}{h_{D2}}}} \cdot b_0 \cdot s_0 \cdot \sqrt{2 \cdot g \cdot h_{D2}} = \psi \cdot b_0 \cdot s_0 \cdot c_{2A}$$

$$= \frac{0.8}{\sqrt{1 + \frac{0.8 \cdot 0.175}{0.36}}} \cdot 2 \cdot 0.175 \cdot \sqrt{2 \cdot 9.81 \cdot 0.36}$$

$$= 0.8 \cdot 2 \cdot 0.175 \cdot 2.255 = 0.631\ \mathrm{m}^3/\mathrm{s}$$

The flow rate is particularly high for the rounded sluice gate plate. The wheel was not overloaded even at the highest flow rates and could safely discharge the water flow. It was therefore of interest to determine the maximum fill factor of the cells:

$$\varepsilon = \frac{Q}{a \cdot u_a \cdot B} = \frac{0.631\ \mathrm{m}^3/\mathrm{s}}{0.35\ \mathrm{m} \cdot 1.534\ \mathrm{m/s} \cdot 2.1\ \mathrm{m}} = 0.56$$

This value closely matches the overload limit identified by Meerwarth of $\varepsilon \leq 0.554$. This overshot waterwheel thus has a specific discharge capacity of $Q_s = 0.300\ \mathrm{m}^3/(\mathrm{s} \cdot \mathrm{m})$. However, during the design process one should not go up to this limit when designing a wheel.

The calculated flow rate is strongly dependent on the contraction coefficient $\psi = h/s_0$ and determines the measurement error for the flow rate very strongly. In this way, the choice of the contraction coefficient greatly influences the efficiency. A measurement of the jet contraction at a submerged sluice must definitely be carried out, but sometimes there is not enough space to measure this parameter safely.

10.3.4 Flow Calculation for the Broad-Crested Weir with Subsequent Chute

The discharge over a weir is calculated using the Poleni's formula. This reads:

$$Q = C \cdot b_0 \cdot h_{ov}^{1.5}$$

The coefficient C depends on the water depth in front of the weir and therefore was determined by matching the data with the WinFlume software [10] (see Fig. 10.24). The Poleni coefficient could be well approximated with a quadratic equation. This results in the following equation:

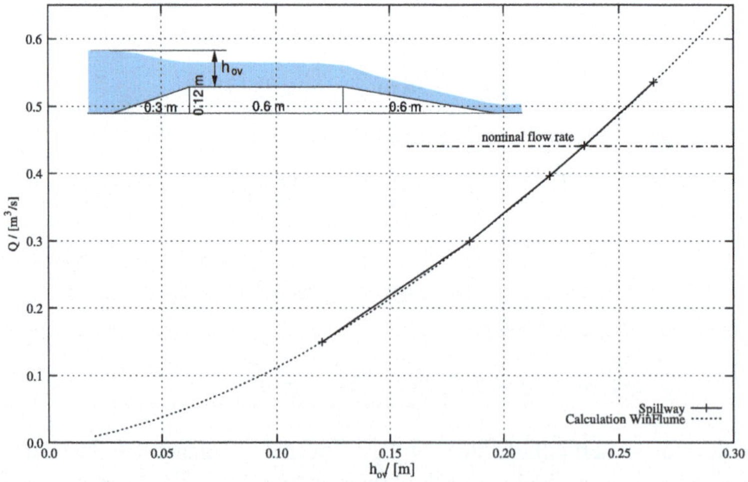

Fig. 10.24 Flow rate versus the overfall height of the broad-crested weir

$$C = a \cdot h_{ov}^2 + b \cdot h_{ov} + c$$

$$a = -1.46006$$

$$b = 1.6933$$

$$c = 1.61774$$

The parameters of the broad crested weir are $b_0 = 2.00$ m, $w = 0.12$ m. The overfall height is defined as:

$$h_{ov}(h_{ow}) = 1.925 - 1.425 - h_{ow}$$

(see Fig. 10.22). The height h_{ow} is the water depth in the headwater.

For example, using the values from measurement point 1 from Table 10.12, the following flow rate is obtained:

$$h_{ov} = 1.925 - 1.425 - h_{ow} = 1.925 - 1.425 - 0.235 = 0.265$$

$$C = -1.46006 \cdot h_{ov}^2 + 1.6933 \cdot h_{ov} + 1.61774 = 1.964$$

$$Q = C \cdot b_0 \cdot h_{ov}^{1.5} = 1.964 \cdot 2 \cdot 0.265^{1.5} = 0.536\ \mathrm{m}^3/\mathrm{s}$$

A compilation of all discharge data for the present overflow can be found in Fig. 10.24. According to the calculations of WinFlume [10], the error in the flow measurement is ±3.6%. This corresponds to approximately ±20 l/s for the maximum value.

The jet at the water was particularly favorable for this arrangement. The design flow rate was also achieved. A jet contraction at the guides of the gates was no longer observable. The discharge was smooth and undisturbed. The noise level was reduced compared to the other measurement setup.

10.3.5 Head Difference, Free Hanging, and Wading

The head difference can simply be determined from the water levels of the staff gauges. Here, according to Fig. 10.20, the following expression applies:

$$H_{geo} = 5.16\ \mathrm{m} - 1.425\ \mathrm{m} - h_{UW} - h_{OW}$$

For example, using the values from line 1 and line 7, column 2 and column 3 from Table 10.11, the following values are obtained:

$H_{geo} = 2.965$ m Opening width of the gates $s_0 = 0.175$ m

$H_{geo} = 3.225$ m Opening width of the sluice $s_0 = 0.055$ m

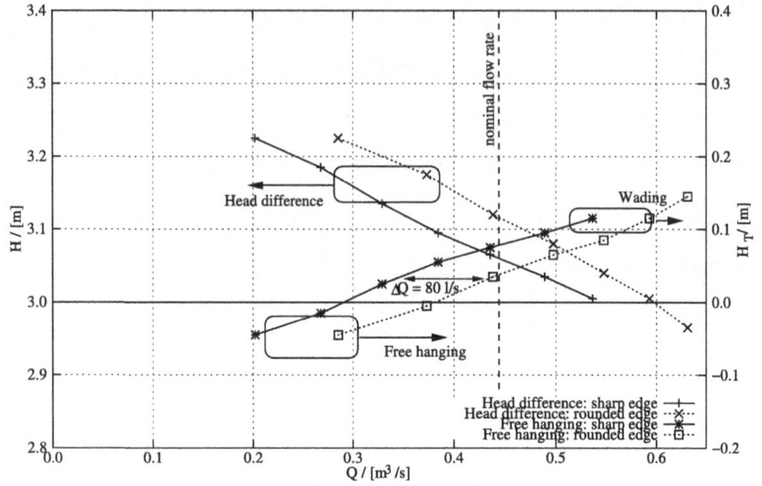

Fig. 10.25 Head difference (left y axis) and free hanging, wading (right y axis) of the overshot waterwheel, plotted versus the flow rate (sluice gates)

In Fig. 10.20, the head difference H_{geo} is plotted against the flow rate Q. This figure shows that there is an offset between the $H_{geo} = f(Q)$ curves for the two sluice gate experiments.

The reason for this is the uneven rubber lip of the sharp-edged sluice. The exiting water jet was not uniform. This results in an offset of $\Delta Q = 80$ l/s. The calculated flow rate is probably too high for the sharp-edged sluice gate. This makes the results of measurement series 1 (sharp-edged sluice) very uncertain.

In Fig. 10.20, it is noticeable that the waterwheel wades at a large opening width. For this reason, the immersion of the waterwheel is shown in Figs. 10.25 and 10.26. The immersion was determined as follows:

$$H_T = -5.16\text{ m} + 4.795\text{ m} + h_{UW}$$

For example, with the values from line 1 and line 7, column 2 and column 3 from Table 10.11, we get:

$$H_T = -5.16\text{ m} + 4.795\text{ m} + 0.51\text{ m} = 0.145\text{ m}$$

$$H_T = -5.16\text{ m} + 4.795\text{ m} + 0.32\text{ m} = -0.045\text{ m}$$

Negative immersion values mean the desired case of free hanging wheel. Positive immersions H_T mean a wading of the waterwheel underwater. The wading starts at a flow rate of $Q = 380$ l/s. Before that, the wheel hangs freely above the underwater.

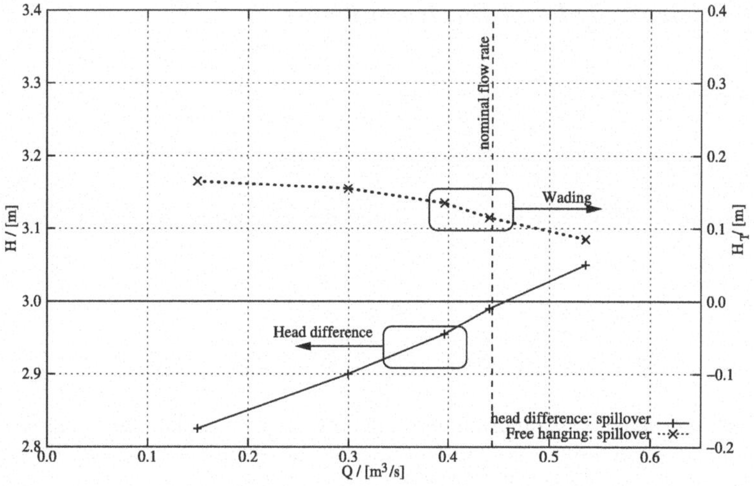

Fig. 10.26 Head difference (left y-axis) and free hanging/wading (right y axis) of the overshot waterwheel, plotted versus the flow rate (broad-crested weir)

The wading of the wheel was also audible. When the system was shut down after a test, one could clearly hear the water flowing into the cells immersed in the tailwater. Observing the wading processes was not possible, as the wheel fits very close to the sidewalls.

It is worth noting that the immersion curves for the measurements with sharp-edged and well-rounded sluices do not align. Since the discharge was completely carried by waterwheel, the water depth in the underwater for a certain flow rate Q must always be the same. A measurement error is excluded here, since the level gauge in the tailwater was fixed throughout the entire measurement period.

In the same way, the head difference and the immersion of the wheel for the experiment with a broad-crested weir have been determined. Since the freewheel sluice gate had to be opened to reduce the overfall height, the tailwater level increased at smaller flow rates above the lower wheel vertex. Therefore, the fall height decreases for smaller flow rates and the wading of the wheel increases. For this reason, the overshot waterwheel did not hang freely above the tailwater for this kind of experiments at all.

This will not occur during normal operation. In practice, it is expected that the underwater will rise as the flow rate increases, while the upper water level also rises. An immersion of the wheel is again expected from a flow rate of $Q = 380$ l/s on. The total usable head difference is smaller in the arrangement with broad crested weir than with the sluice gate. This is because at small flow rates, the upper water level drops with the overfall height, while it rises when using a sluice gate. The tailwater level remains the same for both arrangements with the same flow.

10.3.6 Theoretical and Electrical Power

The theoretical power is defined as:

$$P_{th} = g \cdot \varrho \cdot H_{geo} \cdot Q = 9.8 \cdot H_{geo} \cdot Q \quad \text{in kW}$$

For example, theoretical power is calculated for measurement series 2, measurement no. 1:

$$H = 2.965 \text{ m} \qquad Q = 0.631 \text{ m}^3/\text{s}$$

$$P_{th} = 9.8 \cdot H_{geo} \cdot Q = 9.8 \cdot 2.965 \cdot 0.631 = 18.33 \text{ kW}$$

The electrical power was determined by two methods. The data was read from the wattmeter in the control cabinet. This measuring instrument has an accuracy of ±0.5%.

The error to be considered in this measurement is the reading error, which is higher than ±0.5%. The reading accuracy is 0.2 kW. Based on the value 10 kW, an error of 2% occurs.

The data from column 7 measurement series 1 and 2 are plotted in Fig. 10.27 together with the corresponding theoretical power, which was determined from the measured head difference and the calculated flow rates.

Measurement series 1 (sharp-edged gates) stands out again due to the previously mentioned issue of the calculated flow being too low. The data points for the theoretical power should align on a single characteristic curve.

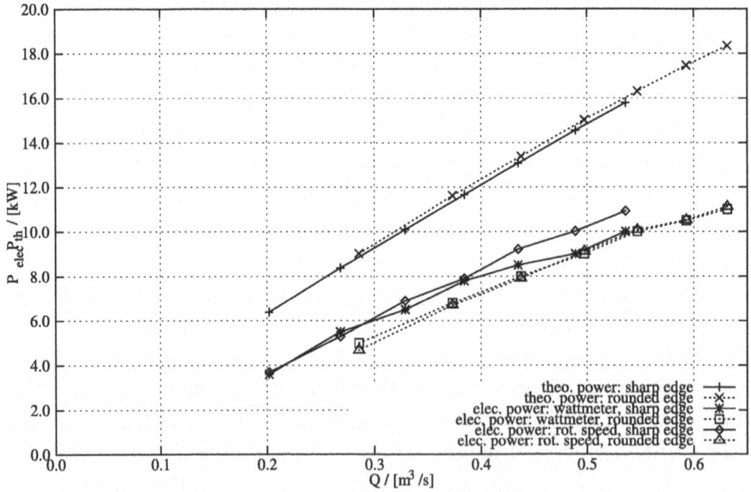

Fig. 10.27 Theoretical and electrical output power, plotted versus the flow rate (Measurement series 1 and 2) sluice gate

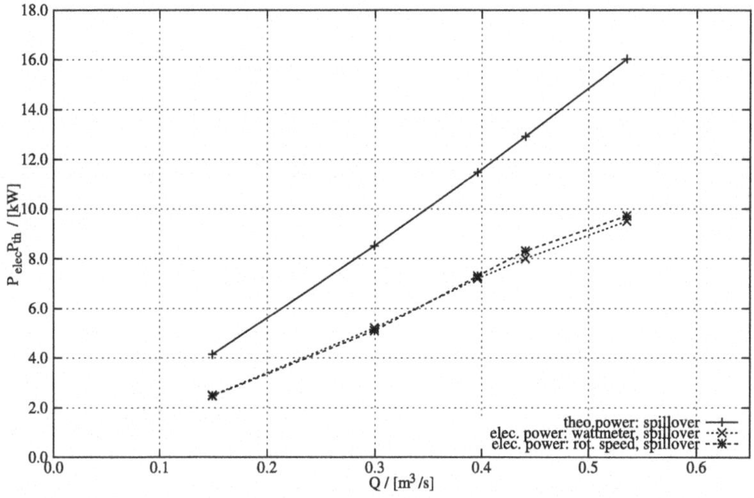

Fig. 10.28 Theoretical and electrical output power, plotted versus the flow of the weir (Measurement series 3)

For the arrangement with broad crested weir, the theoretical and electrical power can be found in Fig. 10.28. This data can be considered reliable.

Furthermore, the speed of the generator was recorded. The specifications of the nameplate did not match the measured speed and the powers (see Fig. 10.29). From the specifications of the type plate, the characteristic equation

$$\begin{aligned} P_{elec} &= \frac{P_{nom}}{n_{nom} - n_{syn}} \cdot (n - n_{syn}) \\ &= \frac{19.5 \text{ kW}}{1032 - 1000} \cdot (n - 1000) \end{aligned}$$

would have been expected. This equation is shown in Fig. 10.29 as a dashed line. The measurement points of measurement series 1, 2, 3 are found as data points. This result was initially very surprising. However, it was then found, by opening the connector plate, that the generator is operated in star instead of delta configuration. The difference between the power-speed characteristics $P_{elec} = f(n)$ results from the Eqs. (7.4) and (7.5).

To circumvent the problem, the electrical readings were plotted against the speed, and the linear characteristic was interpolated using the following equation:

$$P_{elec} = -335.0 + 0.335 \cdot n = -335.0 + 0.335 \cdot f \cdot 60 \tag{10.1}$$

For example, for measurement series 2, measurement point 1, the following value is obtained for the electrical output power of the generator:

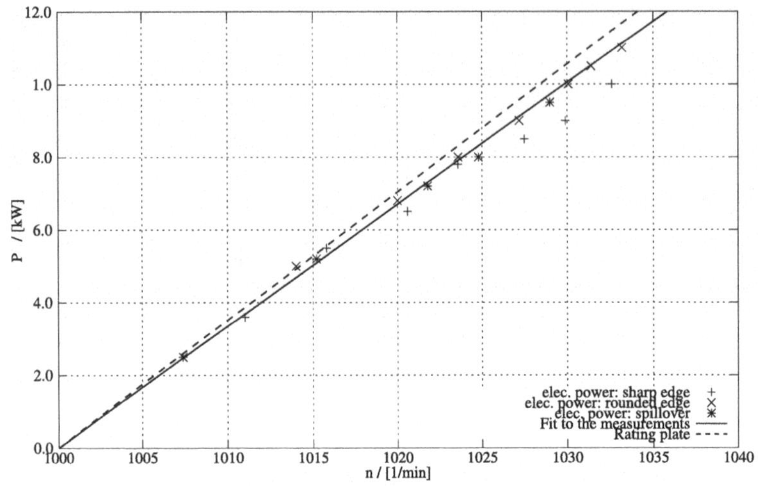

Fig. 10.29 Electrical output power, plotted versus the speed of the generator (Measurement series 1, 2 and 3)

$$P_{elec} = -335.0 + 0.335 \cdot 17.22 \cdot 60 = 11.12 \text{ kW} \tag{10.2}$$

However, this means that the measured values for electrical power measurement from readings of the wattmeter and from the rotational speed measurement were no longer determined independently.

A more precise grid meter reading was not applicable because not all consumers in the network operation were switched off. The results of the current, the voltage and the power factor $\cos\phi$ only provide insufficient results. This example shows that when planning experiments, one should always use several methods to determine the electrical power. Similarly, one should plan different methods to determine the flow rate at a hydropower site.

10.3.7 Determination of the Overall System Efficiency

Reliable measurements for the electrical power were done at the wattmeter and with the speed measurement. Therefore, only these were used to determine the overall system efficiency. The total system efficiency results as the quotient of electrical and theoretical power:

$$\eta_{tot} = \frac{P_{elec}}{P_{th}}$$

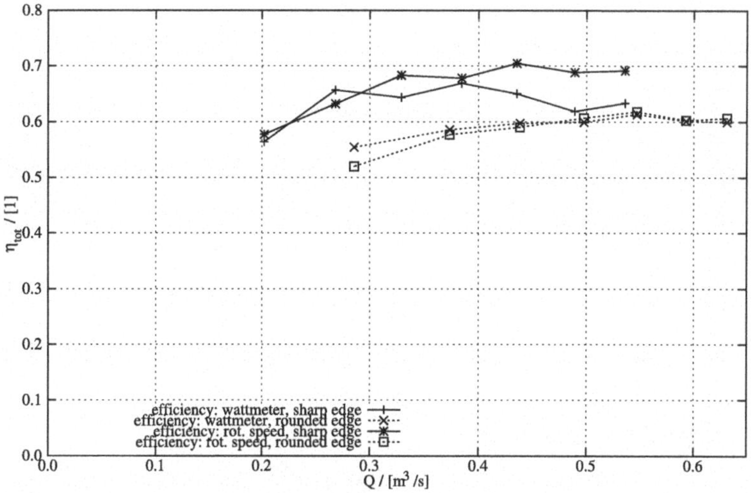

Fig. 10.30 Total efficiency of the overall system (waterwheel/gearbox/generator as well as rake losses), plotted versus the flow rate (measurement series 1 and 2)

The result for measurement series 2, measurement point 1 is as follows:

$$\eta_{tot} = \frac{11.0}{18.33} = 0.600$$

With the presented calculation method, the values in the Figs. 10.30 and 10.31 were calculated for all further measurement series. The results obtained for a sharp-edged sluice are unreliable and yield excessively high efficiencies for our case.

The system efficiency of the plant is consistently high at $\eta_{tot} = 0.6$. Due to the wading and the low efficiency of the generator ($\eta_{gen} = 0.86$), the total efficiency of the plant is not as high as expected from theoretical considerations ($\eta_{tot} = 0.60 \ldots 0.67$).

For this, reference is made to the results from example 9.1 and Fig. 10.16. The data for the broad-crested weir show a similar picture (maximum value $\eta_{tot} = 0.64$). Wading, especially at low flows (compare Fig. 10.26), results in an efficiency drop of 5% compared to the maximum value. The efficiency for the overflow weir is slightly above that of the inflow with rounded sluice, although the waterwheel operates under worse conditions (lower head difference, strong wading at low flow rates).

The reason for this behavior is that the water comes to rest higher in the cells. Due to the smooth inflow in the absence of contraction of the water jet at the guides of the sluice gate, the water does not have to redistribute in the cells towards the side plates.[9] In addition, the head differences and thus the theoretical power are smaller.

[9] This reduces the exit losses.

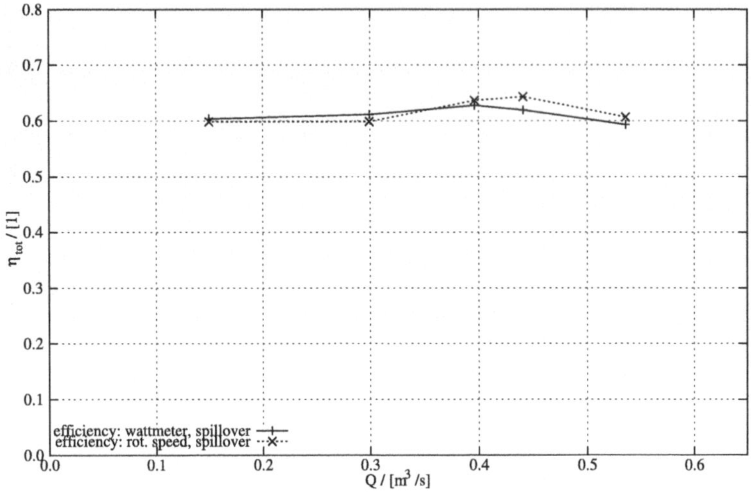

Fig. 10.31 Total efficiency of the overall system (waterwheel/gearbox/generator as well as rake losses), plotted versus the flow (measurement series 3)

Table 10.13 Results of the system efficiency calculation: Measurement series 1 and 2

No	s_0 [m]	h_0 [m]	Q [m^3/s]	H [m]	P_{th} [kW]	$P_{el,W}/P_{el,n}$ [kW]	$\eta_{tot,W}/\eta_{tot,n}$ [1]	Remark
1	0.175	0.370	0.536	3.005	15.79	10.0/10.92	0.633/0.692	Sharp-edged inflow-edge of the sluice gate
2	0.155	0.380	0.489	3.035	14.55	9.0/10.02	0.619/0.688	
3	0.135	0.385	0.435	3.065	13.07	8.5/9.212	0.650/0.705	
4	0.115	0.400	0.384	3.095	11.66	7.8/7.91	0.669/0.678	
5	0.095	0.415	0.329	3.135	10.10	6.5/6.90	0.643/0.683	
6	0.075	0.430	0.268	3.185	8.388	5.5/5.29	0.657/0.632	
7	0.055	0.440	0.202	3.225	6.385	3.6/3.68	0.564/0.577	
1	0.175	0.360	0.631	2.965	18.33	11.0/11.12	0.600/0.606	Rounded inflow-edge of the sluice gate
2	0.155	0.370	0.593	3.005	17.46	10.5/10.52	0.601/0.602	
3	0.135	0.380	0.547	3.040	16.30	10.0/10.08	0.613/0.618	
4	0.115	0.395	0.497	3.080	15.02	9.0/9.11	0.599/0.607	
5	0.095	0.410	0.438	3.120	13.40	8.0/7.91	0.597/0.590	
6	0.075	0.435	0.373	3.175	11.61	6.8/6.70	0.586/0.577	
7	0.055	0.435	0.285	3.225	9.02	5.0/4.69	0.554/0.520	

In Tables 10.13 and 10.14 all determined values for the measurement series 1, 2, and 3 are summarized again. However, the values of measurement series 1 should be viewed with suspicion.

The overall system had an total efficiency of 0.55 to 0.64 in normal operation. The overall system efficiency could be increased by using an efficiency-optimized generator of smaller nominal power and by clearing the tailwater channel by

Table 10.14 Results of the system efficiency calculation: Measurement series 3

No	h_{ue} [m]	Q [m^3/s]	H [m]	P_{th} [kW]	$P_{el,W}/P_{el,n}$ [kW]	$\eta_{tot,W}/\eta_{tot,n}$ [1]	Remarks
1	0.265	0.536	3.050	16.02	9.5/9.715	0.593/0.607	Broad-crested weir with inclinded-chute
2	0.235	0.441	2.990	12.92	8.0/8.308	0.619/0.643	
3	0.220	0.396	2.955	11.47	7.2/7.303	0.628/0.637	
4	0.185	0.299	2.900	8.508	5.2/5.092	0.611/0.599	
5	0.120	0.150	2.825	4.143	2.5/2.479	0.603/0.598	

3...5%. If, for example, a generator with 15 kW and an efficiency of $\eta_{gen} = 0.92$ were used, the system efficiency would be in a range of $\eta?tot = 0.58\ldots0.68$.

If the efficiency of the generator is calculated (see Fig. 10.21):

$$P_{mech} = \frac{21.7}{19.5} \cdot P_{elec} + 0.7 \quad \text{in kW}$$

$$\eta_{gen} = \frac{P_{elec}}{P_{mech}} = \frac{P_{elec}}{\frac{21.7}{19.5} \cdot P_{elec} + 0.7}$$

$$\eta_{mech} = \frac{\eta_{tot}}{\eta_{gen}}$$

from the previous results, a mechanical efficiency of the waterwheel-gear arrangement is obtained. It was assumed that the efficiency of the generator is the same in star and delta connection. The mechanical efficiency is shown in Fig. 10.32. The results (waterwheel and three gear stages) are directly comparable with those of Meerwarth [1] in Fig. 10.10 (waterwheel and two gear stages). The measured arrangement in Stedten has a 10...15% lower efficiency (Figs. 10.33 and 10.34).

This mechanical efficiency is to be expected for cells with bent sheet metal cells, which partly wade underwater. The impact losses at entry due to missing deflection of the jet are high. In addition, the wading loss must also considered.

For the design and dimensioning of the drive of an overshot waterwheel with bent sheet metal blades (see Fig. 8.10), one can assume a waterwheel efficiency of $\eta_{WW} = 0.7\ldots0.76$.

10.3.8 Summary of the Measurement Results

In summary, it can be stated that:

1. The hydropower plant in Stedten has very good conditions for a performance and efficiency measurement, as the design data of the plant is known or could be recalculated well.

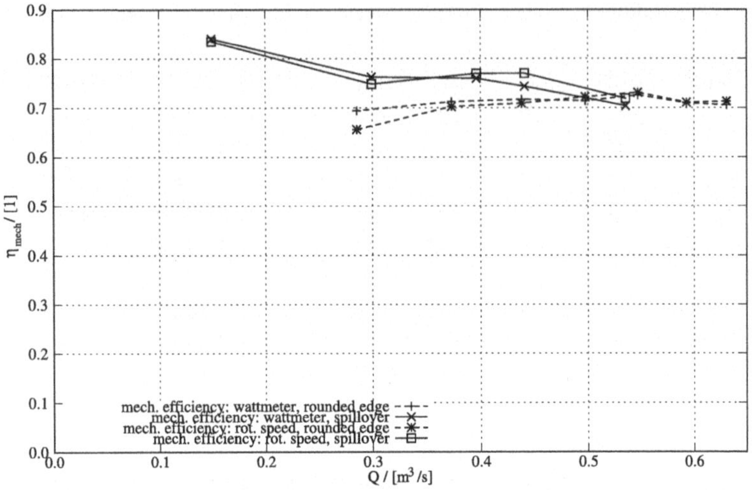

Fig. 10.32 Mechanical efficiency (waterwheel and gearbox), plotted versus the flow rate (only measurement series 2 and measurement series 3)

Fig. 10.33 Weir, fish ladder and sluice gate. The fish ladder is currently overflowed due high flood conditions

2. All measuring devices were available or could be provided. The devices in the metering station were correctly calibrated (which is not always the case with other plants).
3. The measurements taken with a sharp-edged inflow edge of the sluices did not yield meaningful data. This was due to the varying opening width above the

Fig. 10.34 Inlet sluice gate to the diversion channel at the weir

sluice. The likely cause is the sealing of the sluice (rubber lip), which varies in thickness along the length of the inflow edge. It is likely that a water flow $\Delta Q = 80$ l/s or more was exiting from the sluice gate (see Fig. 10.35).

4. The measurements with a rounded inflow edge and with the overfall weir provided comparable results and a consistently high system efficiency of the overall plant.
5. The overall efficiency in normal operation is between $\eta = 0.52$ and $\eta = 0.64$. This efficiency can be increased to about $\eta_{tot} = 0.55 \ldots 0.67$.
6. The overshot waterwheel wades in the tailwater at high flow rates. The wading only led to a surprisingly small decrease in the overall efficiency of the plant compared to the Wisconsin experiments.

The following suggestions can be made to improve the performance of the plant:

1. A further lowering of the tailwater level should be achieved by improving the outflow in the underwater channel since the waterwheel wades in the underwater at high flow rates. The water level should be lowered by $0.05 \ldots 0.1$ m. This is particularly important as the commissioning of the fish ladder is expected to cause an additional water flow of $30 \ldots 50$ l/s.
2. To achieve a high flow rate, the inflow edge of the sluice gate should be well rounded. For this, a semicircular profile made of wood or steel should be used. The radius of the profile should be $R = 0.075 \ldots 0.1$ m.
3. However, it seems better to use a broad crested weir with a subsequent chute. The design of the broad crested weir can be retained if the height of the weir edge (0.12 m) is reduced by 0.02 m (see Fig. 10.24). Under these circumstances,

Fig. 10.35 The overshot waterwheel seen from the tailwater (operation with sharp-edged sluice gate). The contraction of the water jet is noticeable from the guides of the sluice gate towards the center of the jet

the arrangement can supply a flow rate of $Q = 600$ l/s to the waterwheel. In addition, the trash rake at the rake platform could be omitted.

4. The advantage of this arrangement for the operator, who cannot permanently observe the plant, lies in the maintenance-free operation. There is then no need for cumbersome readjustment of the sluice gate to find the optimal power output

Fig. 10.36 Operating platform, trash rake, free wheel sluice gate and sluice gate for supplying water to the waterwheel. On the right is the entrance to the machine house

of the system. The sluice plate is completely pulled out of the water. Only during an emergency shutdown are the holding magnets of the guard activated, which then stop the water supply. The low gradient in the upper water channel is of advantage, as the water level cannot rise so high that the freeboard is overflown. In addition, the fish ladder can be better supplied with water.

5. When using a broad crested weir as a water supply, the inlet structure at the weir in the River Ilm must be automated (flood protection, adjustment of the flow to the system). This achieves better operational behavior and a higher power output of the system.
6. The generator showed a different behavior than specified on the type plate. This was because it was wired in star connection. Replacing the generator with a type with a more efficient type ($\eta_{gen} = 0.92$) and lower nominal power rating would increase the maximum efficiency. The overall system could be increased up to $\eta = 0.67$. The maximum electrical power could be increased to 11.6 kW.

Test series 3 should be repeated once the proposed measures have been implemented. However, the sluice gate of the inlet structure must then be used to control the flow.

On the following pages, some images of the plant in Stedten are compiled. These photographs were taken a few days after the experiment, so the situation described in the text does not always match the images (e.g., flood at the weir) (Figs. 10.36, 10.37, 10.38, 10.39, 10.40, and 10.41).

Fig. 10.37 The generator and on the left side the switch cabinet

Fig. 10.38 Bearing of the waterwheel shaft, coupling and parts of the gear block

Fig. 10.39 The belt drive

Fig. 10.40 Fish ladder following the fish spiral staircase in the form of a pool pass

Fig. 10.41 Fish ladder in the form of a spiral staircase. The water flow was stopped during the measurements

References

1. Meerwarth KW (1935) Experimentelle und theoretische Untersuchungen am oberschlächtigen Wasserrad. PhD thesis, Technische Hochschule Stuttgart
2. Staus A (1923) Maschinenuntersuchungen - I. Band: Hydraulik, 2nd edn. Verlag von Julius Springer, Berlin
3. Staus A (1928) Wasserradversuche. Wasserkraft Wasserwirtschaft 18:259–260
4. Bach Cv (1886) Die Wasserräder, 1st edn. Verlag von Konrad Wittwer, Stuttgart
5. Henne H (1903) Die Wasserräder und Turbinen – ihre Berechnung und Konstruktion. Verlag von Bernh. Friedr. Voigt, Leipzig
6. Nuernbergk DM (2005) Wasserräder mit Kropfgerinne – Berechnungsgrundlagen und neue Erkenntnisse, 1st edn. Verlag Moritz Schäfer, Detmold
7. Weidner CR (1913) Theory and test of an overshot water wheel. Bull Wis Univ Eng Ser Madison, Wisconsin 7(529):157–254
8. Weidner CR (1913) Test of a steel overshot water wheel. Eng News 69(1):38–40
9. Bollrich G, Preißler G (1992) Technische Hydromechanik Band 1, 3rd edn. Verlag für Bauwesen, Berlin
10. Wahl TL, Clemmens AJ, Bos MG, Replogle JA (2013) Winflume, U.S. Department of the Interior, Bureau of Reclamation. www.usbr.gov/pmts/hydraulics_lab/winflume/

Notation

Formula symbol	Designation	Unit
a	Radial crown depth	[m]
A	Area	[m^2]
α_2	Inlet angle	[°]
b	Width in general	[m]
b_0	Channel width/overflow width	[m]
B	Width of the waterwheel	[m]
β	Shape coefficient	[1]
β_2	Blade angle	[°]
c	Speed in general	[$\frac{m}{s}$]
c_0	Inflow speed before the rake	[$\frac{m}{s}$]
c_1	Inflow speed after the rake	[$\frac{m}{s}$]
c_2	Inlet speed	[$\frac{m}{s}$]
c_3	Outlet speed	[$\frac{m}{s}$]
c_4	Speed in the underwater	[$\frac{m}{s}$]
d	Rod thickness	[m]
D_a	Outer diameter	[m]
W	Annual work	[kWh]
ε	Fill factor	[1]
F	Force	[N]
g	Gravitational acceleration	9.81 m/s^2
γ_s	Impact angle	[°]
γ_t	Pitch angle	[°]
η	Efficiency	[1]
hp	Horsepower	[hp]
H	Geodetic head	[m]
H_{Total}	Total/gross head	[m]

D. M. Nuernbergk, *Overshot Waterwheels*, Springer Tracts in Mechanical Engineering, https://doi.org/10.1007/978-3-031-95385-9

Formula symbol	Designation	Unit
H_T	Blade immersion	[m]
h	Water depth in the channel	[m]
h_{max}	Maximum water depth	[m]
h_V	General loss height	[m]
i	Gear ratio	[1]
$\vec{i}$	Unit vector in X-direction	[1]
I	Slope	[1], [m/m]
$\vec{j}$	Unit vector in Y-direction	[1]
k_{st}	Manning-Strickler coefficient	$[m^{1/3}/s]$
$\vec{k}$	Unit vector in Z-direction	[1]
l_u	Wetted perimeter	[m]
m	Weir lowering	[1]
m	Mass in general	[kg]
M	Torque	[Nm]
n	Speed	[1/min]
p	Pressure	$[N/m^2]$
OWS	OberWasserSpiegel= Upper water level	[m.a.s.l.]
P_{elek}	Electrical useful power	[W]
P_N	Mechanical useful power	[W]
P_{th}	Theoretically available power	[W]
Q	Discharge capacity or flow rate	$[\frac{m^3}{s}]$
Q_s	Specific discharge capacity	$[\frac{m^3}{s \cdot m}]$
R_a	Outer radius	[m]
R_i	Inner radius	[m]
R_s	Centroid radius	[m]
r_{hy}	Hydraulic radius	[m]
ϱ	Density of water, $T = 12\,°$	$999.4\ \frac{kg}{m^3}$
s_0	Opening width	[m]
s_K	Tipping slip	[1]
s	Slip	[1]
t_z	Blade pitch	[m]
u_a	Circumferential speed at R_a	$[\frac{m}{s}]$
u_s	Circumferential speed at centroid	$[\frac{m}{s}]$
UWS	UnterWasserSpiegel= lower water level	[m.a.s.l.]
V	Volume	$[m^3]$
W	Work	[Nm]
ω	Angular speed	$[\frac{1}{s}]$
w	Weir height	[m]
X	Coordinate in X-direction	[m]
Y	Coordinate in Y-direction	[m]
z	Number of blades or cells	[1]
Z	Coordinate in Z-direction	[m]
ζ	General loss coefficient	[1]

Index

D. M. Nuernbergk, *Overshot Waterwheels*, Springer Tracts in Mechanical Engineering, https://doi.org/10.1007/978-3-031-95385-9

The manufacturer's authorised representative in the EU is Springer Nature Customer Service Centre GmbH, Europaplatz 3, 69115 Heidelberg, Germany. If you have any concerns regarding our products, please contact ProductSafety@springernature.com

Printed and bound by CPI Group (UK) Ltd, Croydon, CR0 4YY
30/09/2026
02225588-0005